Our Swiss Cheese Universe – A 24-dimensional theory of everything

By

Pardu S. Ponnapalli, Ph.D.

Dedicated to

My son Krishna- may you find all the happiness and joy you deserve in life.

Edition Number: 3U 6/9/2021

- ➢ Added comments about loop quantum gravity. Some grammatical errors, and units of energy and time fixed for uncertainty.
- ➢ Added comments about tie-ins to existing interpretations of quantum mechanics.
- ➢ Added diagram to conclusions and Q&A.
- ➢ Added history of universe section.

Table of Contents

Introduction: Burning Questions in Physics

The current state of theoretical physics is an intriguing and paradoxical one. Fantastic precision and predictions have arisen from the fields of quantum mechanics, general relativity, and the standard model. Yet huge gaps remain in our fundamental understanding of the simplest questions about the grander universe. These include:

- ❖ How can galaxies and galactic clusters stay together when there is insufficient mass to provide the gravitational force that is necessary?
- ❖ Why does General Relativity characterize the universe as a smooth continuous relation between manifold curvature and energy, while Quantum mechanics suggests that discrete bundles of matter and energy is the correct model?
- ❖ Why does matter exhibit wave/particle duality?
- ❖ What constitutes a measurement, the basic mechanism that causes a wave function to collapse into an observable state in quantum mechanics?
- ❖ Why did the universe start with a very slight imbalance between matter and antimatter?
- ❖ Why do we need to use procedures like renormalization (subtracting infinity from infinity to get a renormalized finite value)?

❖ Why is the universe expanding and accelerating in its rate of expansion?

These questions remain unanswered. Most of them fall more into the philosophical interpretation category and seem to have little practical effect on calculations that seem to produce excellent precision. In a way, that makes these questions even more puzzling.

We will examine each of these questions in the subsequent chapters, before proposing a theoretical framework that helps explain all these puzzles. Note that the purpose of the chapters is not to delve deeply into the detail of the puzzles, but rather to pave the way for the new theoretical framework. Readers interested in the details will find references at the end of each chapter.

I have been out of academic physics for nearly thirty years, but it remains my intellectual passion. For all these years, I have been pondering the basic unanswered questions in physics. I have finally arrived at a satisfactory framework that starts to address some of these questions.

This proposed framework will offer answers to some long-standing puzzles and mysteries in physics. It involves understandable structures that obey known physical laws. It conceptually addresses key gaps in our fundamental understanding of the universe.

My motivation for writing this book was to introduce a completely physical framework in which we can address issues raised by physics. Even if some portions of it are inaccessible to us (as is apparent from the uncertainty principle), there should be some logical foundation underneath it.

My vision of the universe involves higher dimensionality. This is nothing unique, as higher dimensions have been postulated by physicists starting a hundred years ago. The great mathematical physicist Kaluza postulated an additional fifth dimension to attempt to unify forces in 1921.

All theories adding dimensions suffer from similar shortcomings. If you want to make it harmonious with relativity, it should have a certain flexibility and dynamism. Since our observations suggest that additional spatial dimensions are not visible to us, we need to suppress them somehow. Techniques to do this are called compactification.

When you start compactifying these dimensions they start becoming unstable. The constant tension between the need for compactification versus the need for dynamism and flexibility leads to a lack of success. The more dimensions that are involved, the more severe the problems .These theories are noteworthy for their mathematical elegance and beauty.

What is unique about the theory I am proposing is the way additional dimensions are introduced. Moreover, the interplay between the additional dimensions and the ones we observe are defined in a novel way. The problem of visibility and dynamism are resolved.

To anyone who doubts the seriousness of the proposal, I am putting a lot of skin in the game. I am offering a substantial $30,000 prize to any qualified physicist who wants to have fun exploring this model and finding a home for it in a peer reviewed journal. Some of the conditions for collecting this prize are outlined in the conclusions.

Chapter 1 Galactic rotation anomalies

To understand galactic rotation anomalies, let us start by looking at our own solar system as an illustrative example of how bodies rotating around a large central mass should behave:

Planet	Orbital period (Earth years)
Mercury	.24
Venus	.62
Earth	1
Mars	1.88
Jupiter	11.86
Saturn	29.46
Uranus	88.01
Neptune	168.8

A clear and logical pattern is apparent. The further you get from the massive central object (the sun) the slower the planet moves. This makes perfect sense even from a Newtonian point of view where the gravity falls off inversely proportional to the distance squared. The further out the object, the weaker the gravitational influence and the slower the planet moves.

Amazingly, this simple reasoning fails for galaxies! The centers of galaxies are now believed to be supermassive black holes with a lot of mass. The stars rotating around them move at the same (or even slightly higher) velocities as you move away from the center. To make the observational reality of the velocity of stars match the theoretical prediction you must assume some additional mass.

This is not the only reason physicists feel the need to postulate the existence of dark matter. Gravitational lensing- where light bends around massive objects led to calculations that showed chunks of mass around galaxies. Dark matter is assumed to form a halo around the galaxies to explain the lensing around distant galaxies.

This proposed dark matter must have some exotic properties including:

> Invisibility (we cannot see it , but it is there influencing gravitational forces)
> Transparency (it does not block our ability to see ordinary matter)
> Capacity to interact gravitationally but not by any other forces (weak, strong, electromagnetic)
> The ability to cause gravitational lensing, the ability to bend light, both at a micro and macro level.
> The ability to affect the universe in different ways in different areas of the universe.

In the late 19th century, physicists were convinced there should be a medium for light to propagate through- they called it aether. There were some compelling reasons to think that there should be such a medium. Michelson and Morley devised an ingenious experiment and tried to measure the effect of aether drag on the speed of light.

Despite several measurements they could not measure either the drag, as the earth moves against the aether. Nor could they measure the acceleration, as the earth moves along with the aether. This experimental result, amongst many other factors, led Einstein to believe there was no aether and the speed of light was constant in a vacuum.

This was one of the founding principles of the Special Theory of Relativity. I think we are in a similar state with dark matter and dark energy. Several ingenious experiments designed to test for the presence of dark matter abound. They are being designed and performed by brilliant scientists all over the world.

Despite all these efforts, there is not a scintilla of experimental evidence that either dark matter or dark energy exist. One might conclude that it would be preferable to find an alternate explanation for the missing mass to explain gravitational anomalies.

An additional mystery surrounding dark matter and dark energy is the apparent lack of any foundation for proportions. Why are observations suggesting a ~5% visible mass, ~25% dark matter and ~70% dark energy split? These proportions seem to have no physical foundation.

Chapter 1: References

1) https://science.nasa.gov/astrophysics/focus-areas/what-is-dark-energy

2) https://astronomy.com/news/2020/03/whats-the-difference-between-dark-matter-and-dark-energy

3) https://www.space.com/20502-dark-matter-universe-mystery-infographic.html

4) https://science.sciencemag.org/content/347/6226/1100.full

5) https://planetseducation.com/dark-matter-and-dark-energy/

Chapter 2: General Relativity vs Quantum Mechanics

For many years, the contrasting views of General Relativity and Quantum Mechanics have been apparent to physicists. General Relativity paints a picture of a smooth continuous universe, with mass perturbing the space time fabric. Moreover, a fundamental tenet in relativity is that space and time are one combined dimension.

As Minkowski famously proclaimed:

Henceforth space by itself, and time by itself, are doomed to fade away into mere shadows, and only a kind of union of the two will preserve an independent reality."

In contrast Quantum Mechanics paints a picture of the universe composed of discrete packets of energy. Assumptions underlying quantum mechanics lead to the concept of matter that simultaneously behave like waves and particles, the uncertainty principle, and other mind-bending results.

If you want to get an intuitive grasp of this without delving deep into the math, think about a peg board as the quantum mechanical view of the universe, and a smooth sheet of paper as the relativistic view.

Armed with a pencil, you can traverse the sheet of paper smoothly without ever lifting it, a characteristic of a continuous object. However, in the peg board with holes, you must jump from hole to hole in discrete jumps- there is no continuous motion.

One of the intriguing features of these models is that both are fantastically accurate and produce stunning predictions about the behavior of the universe that are experimentally verified. If you want to look at a practical application of these theories, you need not look any further than the mobile phone you are likely holding right now.

The semiconductors and transistor technology that makes the phone operational is based on quantum mechanics and quantum field theory, while the GPS feature it uses works only because special relativistic and general relativistic effects are considered.

It is intuitively obvious that both theories cannot be simultaneously true- the universe cannot be granular and continuous at the same time. This becomes apparent when you try to unify quantum mechanics and gravity- it leads to theories that end up with infinite quantities that cannot be removed. These theories are known as non-renormalizable.

We will touch upon renormalization in a later chapter. Ideally, we would have a theory that very naturally suggests granularity at small scales and continuity at large scales- this is no small task, if you will forgive the pun.

Chapter 2: References

1) https://www.infoplease.com/math-science/space/universe/theories-of-the-universe-quantum-mechanics-vs-general-relativity

2) https://science.howstuffworks.com/science-vs-myth/everyday-myths/unite-quantum-mechanics-general-relativity.htm

3) https://www.azoquantum.com/Article.aspx?ArticleID=83

4) http://scivenue.com/2017/11/16/quantum-mechanics-general-relativity/

5) https://now.northropgrumman.com/when-quantum-mechanics-and-relativity-collide/

Chapter 3: Wave/Particle Duality

For a long time, physicists pondered the nature of a photon: Was it a particle, like a billiard ball with solid content and predictable motion, or was it a wave like the wave on the surface of the ocean? Intriguingly, experimental evidence pointed to both conclusions!

Eventually, the quantum mechanical interpretation of this was that it is both a particle and a wave. Or to put it another way, photons exhibit the properties of particles in some situations, and a wave in others.

To say this is counterintuitive is to understate the matter. It makes it difficult to visualize what a photon is. You must just follow the math as prescribed by quantum mechanics and accept the results. To fully illustrate this quixotic behavior of the photon, consider the following properties of light:

> Reflection: This can be explained by either the wave or particle view. If you think of a photon as a billiard ball, reflection amounts to it bouncing back to you directly in the line it moved forward in. If you think of a collection of photons as a wave, it amounts to crashing into a barrier and traversing back in the same form.

> Refraction: Changing the direction of a wave as it passes to another medium seems intuitively feasible. A particle bending this way- not so much. Newton did have an ingenious explanation using the particle theory of light for refraction, but it did not stand the test of time. So, this phenomenon tilts us to the wave interpretation.

➢ Interference: Light patterns of different wavelengths can interfere with each other and produce a different output. It is easy to see how two waves add up or nullify each other . Not so much with particles, so this again tilts us to the wave interpretation.

➢ Photoelectric effect: When light hits a metallic surface, electrons are ejected from it. But it does so in an extremely specific way. There is a threshold frequency of the light that must be exceeded for the phenomenon to occur. Below that frequency, no matter how intense the light, there is no effect. And there is no correlation between the intensity of the light and the maximum energy of the electrons. All these facts taken together could only be explained by assuming light was particle-like. This tilts us to the particle interpretation.

So, the final answer seems to be almost mystical : Sometimes light behaves as a wave and sometimes as a particle, with no underlying indication when it decides it wants to be a wave and when it decides it wants to be a particle.

It reminds you of the quote from Walter M. Miller, Jr., "...in divinity opposites are always reconciled." Or the popular jingle: "Sometimes you feel like a nut, sometimes you don't." Apparently, for light "Sometimes you feel like a wave, sometimes you don't".

Chapter 3: References

1) https://byjus.com/jee/photoelectric-effect/#history-of-the-photoelectric-effect
2) https://www.thoughtco.com/wave-particle-duality-2699037
3) https://opentextbc.ca/universityphysicsv3openstax/chapter/wave-particle-duality/
4) https://alevelphysics.co.uk/notes/wave-particle-duality/
5) https://science.howstuffworks.com/light6.htm

Chapter 4: The Uncertainty Principle and the Measurement problem

There are various ways that the measurement problem is stated, but the core idea is that the observed physical state of the system is determined by a collapsing wavefunction. The wavefunction itself is a superposition of several states in an extremely complicated mathematical space with probability coefficients.

The uncertainty principle has two main forms. The first one considers energy and time as conjugate pairs that cannot be simultaneously determined with great accuracy. The second one considers position and momentum as a conjugate pair and asserts similar conclusions.

Let us look at the case of one dimension in space and an associated momentum in that direction. The uncertainty principle states that:

$$\Delta x \, \Delta p_x \geq \hbar/2 \qquad\qquad [4.1]$$

$\hbar$ is Planck's constant, which is an extremely small number on the order of 10^{-34} J-s (units of energy times time). So, the more precisely we figure out the position, the smaller Δx becomes and the larger Δp_x becomes. And conversely, the more precisely we determine the momentum, the larger the uncertainty in the position. Pinning down a particle is like chasing a ghost in quantum mechanics. If you figure out how fast it is moving with precision, you lose track of where it is. Conversely, if you pin down where it is, you have little idea how fast it is moving.

Let us look at the energy-time equation:

$$\Delta E \, \Delta t \geq \hbar/2 \qquad\qquad [4.2]$$

As the clock starts ticking, we can zoom in on the slice of time and consider a starting point and an endpoint t_1 and t_2. The more precisely we determine the tick and a tock of this interval, the less clear we are on the energy of the object. Conversely, the clearer we are about the energy, the more clueless we get about the tick and the tock of the particle existence.

The uncertainty principle raises a lot of philosophical questions. First, this is not what we expect based on our macroscopic experiences; I can track the precise position of a billiard ball and its momentum with no problem. Since the billiard ball is composed of a large number of atoms which obey quantum mechanics, this seems like an obvious contradiction. Secondly, this idea is a consequence of the notion that a particle like an electron is a probability wave that collapses into a specific physical state when you take a measurement. This odd behavior is kind of inexplicable, and since microscopic quantum events are tied to macroscopic events , it leads us to paradoxical situations. Thirdly, it raises the whole question of what constitutes a measurement. Does a human being have to measure a state for it to count? Can a robot be involved in measurements? What if you can measure in different ways?

A classic paradox that is raised by this is called Schrödinger's Cat Paradox. I will not delve into it here, but I provided some references for the curious reader.

Chapter 4: References

1) https://byjus.com/jee/heisenberg-uncertainty-principle/
2) https://www.ted.com/talks/chad_orzel_what_is_the_heisenberg_uncertainty_principle/transcript

3) https://exploringyourmind.com/all-about-heisenbergs-uncertainty-principle/

4) https://www.livescience.com/23426-uncertainty-principle-measurement-disturbance.html

5) https://www.thoughtco.com/what-is-schrodingers-cat-2699362

Chapter 5: Matter Antimatter Imbalance

If the universe started as a singularity as the Big Bang postulates, then the energy should have split into an even amount of matter and antimatter. Unfortunately, our observations are that antimatter is extremely rare and can be produced only in giant apparatus at research institutions like CERN.

Antimatter has the same mass as their matter counterparts but have opposite quantum properties like charge and spin. When matter meets antimatter, the result is total annihilation and creation of pure energy. This is the purest form of creating energy and used as a propulsion method in my favorite science fiction show "Star Trek".

In our universe, somehow more matter survived than antimatter. This is ascribed to some perturbation that manages to tilt the process in favor of matter- which is an extremely sensible proposition. Unfortunately, no experimental evidence has been gathered to support such a proposition.

Consequently, the matter-antimatter imbalance has no explanation. Physicists have looked hard for an explanation- many researchers have come up with theories about this broken symmetry. Sadly, none have panned out. Basically, all observations support the proposition that if the Big Bang theory is correct, our universe should be made of 50% matter and 50% antimatter.

The standard model , which explains several observations about the behavior of elementary particles, is one of the most successful theories in physics. It accounts for the behavior and properties of quarks, electrons, the Higgs Boson, and their interactions. But it comes up dry when it comes to an explanation of matter antimatter imbalance in the universe.

When you have problems explaining the very existence of the universe, it might be time to start looking for some radically different fundamental ideas.

Chapter 5: References

1) https://weirdnews.info/2021/03/21/the-universe-shouldnt-exist-cern-scientists-announce/
2) https://home.cern/science/physics/matter-antimatter-asymmetry-problem
3) https://scitechdaily.com/imbalance-between-matter-and-antimatter-helps-explain-our-existence/
4) https://phys.org/news/2019-02-universe_1.htm
5) https://www.sciencetimes.com/articles/29814/20210222/discovery-quantum-theory-solve-baryon-asymmetry-problem.htm

Chapter 6 Renormalization

It is a feature of quantum field theory that a renormalization takes place in the equations. Two infinite factors are subtracted to give rise to a finite factor. There is little argument that it works brilliantly and produces agreement between theory and experiment to the tune of one part in several billions. Here is what two of the most brilliant physicists in history had to say about renormalization from WikiQuote (quotes are only partially reproduced):

- "Sensible mathematics involves neglecting a quantity when it turns out to be small—not neglecting it just because it is infinitely great and you do not want it!"

P. A. M. Dirac, *Directions in Physics* (1978), 2. Quantum Electrodynamics

- "What is certain is that we do not have a good mathematical way to describe the theory of quantum electrodynamics: such a bunch of words to describe the connection between n and j and m and e is not good mathematics."

Richard Feynman, *QED: The Strange Theory of Light and Matter* (1985), Chap. 8. Loose Ends

The question is how do you interpret this bizarre process? There is no mathematical justification for it beyond the fact that it works and the physical foundation for it seems murky. Renormalization involves numerous processes that sound like they are science fiction.

The first is when a photon creates a virtual electron-positron pair that annihilate each other. The question is, what is the source of this "virtual" pair? That question is deemed irrelevant since the numbers all work out precisely in the end.

The second scenario is when an electron emits and absorbs a virtual photon, leading to a self-energy condition. Once again, why is the photon virtual? What does that mean?

Why does the self-energy turn out to be infinite and need to be subtracted out? All these questions are again deemed of academic interest. If you approach a theoretical physicist with these questions, you may get a shrug of the shoulders, followed by a return question: Who cares ?

The third scenario is where an electron emits a photon, then emits a second one, and subsequently absorbs the first one. Once again, truly little rationale is offered, since the numbers all work out.

As bizarre as all this seems, the result is spectacular. Calculations of the Lamb shift, which is the energy difference between two quantum states in the hydrogen atom, based on quantum electrodynamics and renormalization are mind bogglingly precise.

Those measured experimental values of properties of the electron match calculated values to an extremely high degree of accuracy. This leaves no doubt that however messy the process, there is profound truth buried somewhere in the renormalization process.

Chapter 6: References

1) https://en.wikipedia.org/wiki/Renormalization
2) https://www.preposterousuniverse.com/blog/2013/06/20/how-quantum-field-theory-becomes-effective/
3) http://www.physics.umd.edu/courses/Phys851/Luty/notes/renorm.pdf (this one is heavy on the math)
4) https://arxiv.org/pdf/hep-th/0212049.pdf (this is also heavy on the math; exceedingly difficult to avoid when delving into renormalization)
5) https://www.tokenrock.com/explain-quantum-electrodynamics-38.html

Chapter 7: Our Expanding Universe

Why does the Universe expand? You would think it is a matter that has been resolved in cosmology. Einstein introduced a constant in his equations, the cosmological constant, to counteract gravitational forces and keep it static. But after numerous observational results, it was clear that not only was the universe expanding it was doing so at an accelerating rate.

Imagine a balloon, whose surface area is increasing as air is pumped into it. If it expanded at a steady rate of say 1 inch increase in radius per 1 minute, then you would have a volume 8 times the original after 1 minute (assume the original radius is 1 inch and the balloon is a perfect sphere). After 2 minutes , the radius would now be a total of 3 inches and the volume would be 27 times the original. This would be like a steadily expanding universe.

Take the same balloon and assume that the rate itself is accelerating. In the first minute the radius goes up by 1 inch. But in the second minute the radius goes up by 2 inches and in the third minute the radius goes by 3 inches. After 1 minute , you have a radius of 2 inches total and a volume 8 times the original.

After 2 minutes, since the rate has gone up, the radius is a total of 4 inches , resulting in 64 times the original volume. After 3 minutes, the total radius is 7 inches, and the volume is 343 times the original. You can see the dramatic effect on size of an accelerating rate of expansion.

As it turns out, our universe is very much like the second balloon and Einstein's cosmological constant, which he regarded as a blunder, explains the expansion! This is an ad hoc constant though, and no one knows what the underlying force is that is driving this expansion.

The force is called dark energy as a placeholder, but that does little to explain either its properties, origin, or where it exists. Not only is it supposedly responsible for the expansion of the universe, but it also constitutes most of all energy in the universe.

If you thought science fiction was weird, this should make you realize that cosmology is a close competitor. This is as mystical a force as "The Force" in Star Wars.

This is no doubt one of the deepest mysteries of the cosmos.

Chapter 7: References

1) https://www.space.com/52-the-expanding-universe-from-the-big-bang-to-today.html
2) https://phys.org/news/2021-03-fast-universe-galaxies.html
3) https://themidnightsky.com/our-expanding-universe/
4) https://www.nasa.gov/feature/goddard/2019/mystery-of-the-universe-s-expansion-rate-widens-with-new-hubble-data
5) https://www.space.com/universe-expanding-fast-new-physics.html

Chapter 8 Our Swiss Cheese Universe explains it all.

By way of summary, our current state of theoretical physics embraces:

- ➤ Sophisticated, accurate models that cannot explain why the universe must be smooth and discrete simultaneously.
- ➤ Models that postulate that matter is made of "wavicles" – things that exhibit properties of waves and particles at the same time.
- ➤ Models that cannot even explain why we came into existence at all due to the slight imbalance of matter and antimatter.
- ➤ Models that cannot tell us what even a measurement means in a clear manner.
- ➤ Models that make you subtract infinity from infinity and end up with a finite number.
- ➤ Equations for fundamental particles that are so complex that only a select few on the planet even understand them.
- ➤ Finally, models that cannot explain the expanding nature of our universe, nor the very fundamental question of energy in the vacuum.

Obviously, this is not an ideal situation. Most of these problems are dismissed with the refrain that underlying theories "work" meaning they individually produce very precise results. I think what is required is a conceptual framework that swings for the fences; there is no point in picking at the margins.

At the risk of sounding ambitious, you need a framework that clears up all the major problems in one blow. This framework should preserve the ideas of all the different branches and reduce to their equations in limits that pertain.

For example, it should be a natural outcome of the framework that going to quantum levels implies different behavior of particles, or that going to massive levels makes you work in a general relativistic frame. You should be able to deduce this in a straightforward way from the fundamentals of the framework, without having to do extreme mental gymnastics or get lost in a tree of detailed mathematics and losing sight of the forest.

Surprisingly, I think such a framework exists. Let us look at two fundamental facts:

> To conform to a relativistic view of the universe, you must view space and time as one dimension.
> A substantial number of quantum mechanical based quantum field theories adopt "hidden dimensions" to move towards unified field theories.

To set up the new framework, we must combine the best ideas of relativity and quantum mechanics (and their evolution into quantum field theories and unified field theories). This leads us to the following basis for our new theory:

> In all treatments of physics, we will always treat space and time as a combined dimension.
> There are hidden dimensions as many unified field theories predict.

Additionally, based on our discussion of existing mysteries of current theories, we can arrive at the following guidance:

- ➤ There is no broken symmetry that caused a matter antimatter imbalance in our universe. There is no experimental evidence for this imbalance and our new theory should suggest that somehow it is corrected.
- ➤ Dark matter and Dark energy are just placeholders in our bookkeeping. We need to account for this matter and energy in some systematic way. In this effort, we need to explain galactic rotational anomalies in an understandable manner.
- ➤ The theory needs to flow smoothly from the macroscopic to the microscopic. Quantum laws should suggest themselves at the microscopic level, while our view of a smooth, continuous universe at large scales should be reinforced.
- ➤ Renormalization needs to somehow involve real energies without infinities. The "virtual" processes in renormalization need an explanation.
- ➤ There should be some underlying explanation of what constitutes the uncertainty principle and gives insight into why the universe becomes so "fuzzy" and probabilistic at quantum levels. Our new theory should help us out with this.
- ➤ The new theory should shed light on the fundamental quantum property of wave/particle duality, identifying an underlying reason for this quixotic behavior.
- ➤ Finally, the new theory should explain why the cosmological constant in Einstein's field equations needs to be there to

account for an expanding universe that is accelerating its expansion.

What type of model would accommodate this type of framework? I propose the following:

- The original primordial universe was composed of multiple spacetime dimensions, with a total of 24 dimensions. The correct way to look at this, in the spirit of relativity, is as 6 spacetime manifolds, each composed of 4 dimensions.
- After the Big Bang, it split into separate four-dimensional spacetime manifolds (in other words the hidden dimensions of unified field theories are retained but they are hidden spacetime dimensions). We live in one of the fractured spacetime manifolds that are interweaved into the larger composite spacetime manifolds.
- Three of these cleaved manifolds are dominated by matter , while the other three are dominated by antimatter. As the original primordial universe was pure energy, the amounts exactly balance each other.
- For reasons that will become apparent, the superstructure has 6 four dimensional spacetime manifolds (24 total space and time dimensions if you want to count them separately). Note that the total is remarkably close to some superstring theories with 26 dimensions but is distributed more symmetrically, without any bias towards extra spatial dimensions only.

- Separating each spacetime manifold 1,2,3,4,5 and 6 is a layer of energy density (this is the energy that is currently called dark energy)
- Each spacetime manifold has small folds, which I will call quantum folds. The fold is of such a nature that quantum motion involves stepping through a cycle of all the manifolds: 1->2->3->4->5->6->5->4->3->2->1.
- Each spacetime manifold has ordinary matter in it, but due to the separation of the manifolds and the inter-manifold energy density, each manifold identifies the matter in all other manifolds as "dark matter".

To set this framework up, let us travel back to the time of the Big Bang and set up a thought experiment. Imagine that the universe was a small three-dimensional slab at the time of creation. It has great extent in x and y directions but truly little extent in the z direction.

Now a massive explosion shears the universe and expands it so that you are left with six adjacent slabs. Each slab would look like an ordinary plane. The slabs are connected by an "interslab" material that inhibits most interactions on a macroscopic scale but have microscopic holes in them that lets particles move between slabs if they are small enough.

Living in one slab, you could systematically use what you think is a very logical method and arrive at an exceptionally low number for the original mass of the universe. It would result in apparently contradictory laws because a large amount of the mass is simply inaccessible to you yet seems to affect your world.

Others who lived in slabs 2 to slab 6 would arrive at similar conclusions. You might be tempted to label the missing mass "dark matter". Physical laws you develop would only describe motion along the x and y axes in slab 1. They would produce accurate predictions for your day -to- day purposes.

You might notice that massive particles seem to behave differently, because they bend your slab so much, they affect other slabs, and matter from other slabs affect them. Also, you will observe that the actual motion of particles seems to be affected by some 'invisible' matter at large scales. And finally, you see that smaller particles exhibit some bizarre behavior (since they slip between slabs through microscopic holes).

In such a hypothetical universe, the only way to get a complete view of the total physical universe is to combine the views of the people from each slab and interweave them to form the known universe.

I presented this simplistic thought experiment with physical dimensions as a segue into my proposed model. The repeated and accurate theme of our Einsteinian universe is that the fundamental dimension is spacetime. So instead of our imaginary three-dimensional slab that sheared into different shards at the time of the Big Bang, the original primordial universe was compressed into one 24 dimensional spacetime manifold in a very tight compact space.

All the energy of the universe would be concentrated there. Very shortly after the beginning, it split into six separate four dimensional spacetime manifolds, much as the three-dimensional slab split into six different ones. We live in one such spacetime manifold.

The number six is arrived at for the split judging by the estimated amount of mass missing to explain gravitational phenomena. We live in one of the fractured spacetime manifolds that are interweaved into the larger 24-dimensional universe. The uniformly distributed mass gradually spread out over the six space time manifolds after the original Big Bang.

Each spacetime manifold views all the mass in other manifolds as "dark matter". The matter in these other manifolds causes observable effects in the 'native' manifold. The universe is built out of a 24-dimensional superstructure of which we inhabit a 4-dimensional space time manifold. At a microscopic level, spacetime is quantized. A movement from point A to point B, is a movement through all 6 spacetime manifolds. I will supply more mathematical detail on this notion in due course but the basic coordinates in spacetime appear as follows:

- Coordinates in spacetime manifold 1: ((x^0, x^1, x^2, x^3), 0, 0, 0)
- Coordinates in spacetime manifold 2: (0, (x^4, x^5, x^6, x^7), 0, 0, 0, 0)
- Coordinates in spacetime manifold 3: (0, 0, $(x^8, x^9, x^{10}, x^{11})$, 0, 0, 0)
- Coordinates in spacetime manifold 4: (0, 0, 0, $(x^{12}, x^{13}, x^{14}, x^{15})$, 0, 0)

- Coordinates in spacetime manifold 5: $(0, 0, 0, 0, (x^{16}, x^{17}, x^{18}, x^{19}), 0)$
- Coordinates in spacetime manifold 6: $(0, 0, 0, 0, 0, (x^{20}, x^{21}, x^{22}, x^{23}))$

This abbreviated coordinate representation puts a single zero for quadruple coordinates where appropriate. Note that each individual spacetime manifold is a submanifold of the overall 24-dimensional universe. The 24-dimensional universe is a superposition of 6 individual spacetime manifolds, each with 4 dimensions.

They are separated by intermanifold energy which is what we label as dark energy. At a quantum level, particles transition from spacetime 1 to 2 to 3 to 4 to 5 to 6 and then backwards from 6 to 5 to 4 to 3 to 2 to 1. The separation between transitions like this defines the fundamental quantum fold of spacetime.

Note that there is a "shadow displacement" in our native manifold while these transitions take place. As the quantum particle cycles from 1->2-3->4->5->6->5->4->3->2->1, there is an apparent displacement in manifold 1 that looks like the particle disappeared from where it started and reappeared elsewhere. All that "time" is spent in manifolds 2,3,4,5, and 6.

This might sound like an odd method of motion but in fact it conforms to exactly what quantum mechanics predicts for movement. Particles move from point A to point B with no evidence that there was any movement in between.

Spacetime is quantized at such a small scale that it looks continuous at anything beyond quantum scales. When you get down to quantum scales the continuous transitions between the manifolds result in the bizarre behavior found in quantum particles.

The original universe is a direct sum of the 6 spacetime manifolds. If we let the universal manifold be denoted M and each manifold as M_i, then :

$$M = M_1 \oplus M_2 \oplus M_3 \oplus M_4 \oplus M_5 \oplus M_6$$

The overall manifold is a direct sum of submanifolds (not an ordinary addition). The notion of geometry and physical displacement is quite simple in this primordial universe. For example,

$ds_1^2 = dx_0^2 - dx_1^2 - dx_2^2 - dx_3^2$ *(spacetime manifold 1)*
[8.1]
$ds_2^2 = dx_4^2 - dx_5^2 - dx_6^2 - dx_7^2$ *(spacetime manifold 2)*   [8.2]
$ds_3^2 = dx_8^2 - dx_9^2 - dx_{10}^2 - dx_{11}^2$ *(spacetime manifold 3)*   [8.3]
$ds_4^2 = dx_{12}^2 - dx_{13}^2 - dx_{14}^2 - dx_{15}^2$ *(spacetime manifold 4)*   [8.4]
$ds_5^2 = dx_{16}^2 - dx_{17}^2 - dx_{18}^2 - dx_{19}^2$ *(spacetime manifold 5)*   [8.5]
$ds_6^2 = dx_{20}^2 - dx_{21}^2 - dx_{22}^2 - dx_{23}^2$ *(spacetime manifold 6)*   [8.6]

Note that here $dx_0^2 = c^2 (dt_0)^2$, $dx_4^2 = c^2 (dt_1)^2$, $dx_8^2 = c^2 (dt_2)^2$, $dx_{12}^2 = c^2 (dt_3)^2$, $dx_{16}^2 = c^2 (dt_4)^2$ and $dx_{20}^2 = c^2 (dt_5)^2$. Any general displacement is a direct sum of these, not an ordinary addition.

$$ds^2 = ds_1^2 \oplus ds_5^2 \oplus ds_3^2 \oplus ds_4^2 \oplus ds_5^2 \oplus ds_6^2 \quad [8.7]$$

This universe is like having 6 distinct Einsteinian spacetime manifolds that coexist together to form a 24-dimensional universe overall. Within each manifold all our "regular" laws apply. You can just take all our physical laws and replace each coordinate subscript with the subscript +4 and write the equations.

To imagine what 'movement' would look like in this primordial universe, imagine that the 6 distinct spacetime manifolds correspond to 6 distinct hiking trails. The composition of the trails is as follows:

- ➤ Trail 1 : dirt
- ➤ Trail 2: paved asphalt
- ➤ Trail 3: concrete
- ➤ Trail 4: grass
- ➤ Trail 5: sand
- ➤ Trail 6: rubber.

You start walking trail 1 on the dirt. You travel for about 1 mile in 15 minutes, and suddenly you find yourself on a paved asphalt trail traveling say 2 miles in 30 minutes. Since your consciousness is geared to the notion of 6 distinct spacetime manifolds, you find nothing strange in this.

 Your internal clock accepts that you must start a separate timer for this path, and you are aware you traveled for 30 minutes on this track. This behavior repeats for trail 3-6 with say 3 miles/45 minutes, 4 miles 1 hour, 5 miles 1.25 hours and 6 miles 1.5 hours.

 You end up on a rubber hiking trail with no relation to where you started. Because you are intrinsically aware of the nature of your universe, you understand that your overall displacement was .0000054 light-seconds on hiking trail 1 in what you could tell was a distinct period of 15 minutes in submanifold 1, 0000107 light seconds on hiking trail 2 in what you could tell was 30 minutes on hiking trail 2 , etc.

Your perception of light and how far it travels in each submanifold is tied to a different chronological drummer. But the speed of light is the same in each submanifold. This is a very strange state indeed, and maybe it is fortunate the universe did not stay in this state, since this does not conform to our observation of it, and it might take a madman to appreciate how it would work.

The caveat to the notion of the madman appreciating this universe is that none of these notions are all that strange in the quantum world; in fact, they are tame compared to ideas like quantum entanglement.

The Big Bang fractured this universe and cleaved it into 6 separate space time manifolds which are connected only by quantum folds and separated by a layer of energy density. Movement in this universe conforms to our observations in our resident manifold. A general displacement for smaller scale particles is now through quantum folds in each of the submanifolds and then a return to the original manifold.

For larger scale particles the displacement is exactly as we expect using our laws of relativity. Going back to our hiking analogy, your walk is confined to trail 1 for 1 mile in 15 minutes. If you could somehow slow down your steps so finely so that they are at the scale of the quantum folds you would have a flash of awareness of being in manifold 2 , then 3, then 4, then 5 , then 6.

Then you would head back through 5 , 4, 3, 2 and finally returning to trail 1. Without the ability to granulate your steps so finely, you would sense only that you are moving continuously on trail 1.

There is a shadow displacement through all those trails for an extraordinarily small slice of space time. So small that it makes no difference to the way you measure distances on a macroscopic scale. On a quantum scale these flows through the other trails are a key feature in displacement and make the laws of physics look bizarre.

To visualize the cleaving of the space time manifolds, let us start with a simple model of viewing the Big Bang singularity as a point. If it expands after an explosion, the most logical expansion would be to a circle. Imagine the circle as a clock face with a regular 12-hour clock with thick bold markings separating every 2 hours (note, not every hour).

Think of every 2-hour segment of the clock as a manifold . Manifold 1 is noon to 2 pm, manifold 2 is 2 pm to 4 pm, etc. Between the segments would be the bold markings representing the intermanifold layer. There is some intuitive sense in a point exploding radially out and creating a circle in equal segments. This is pictured below:

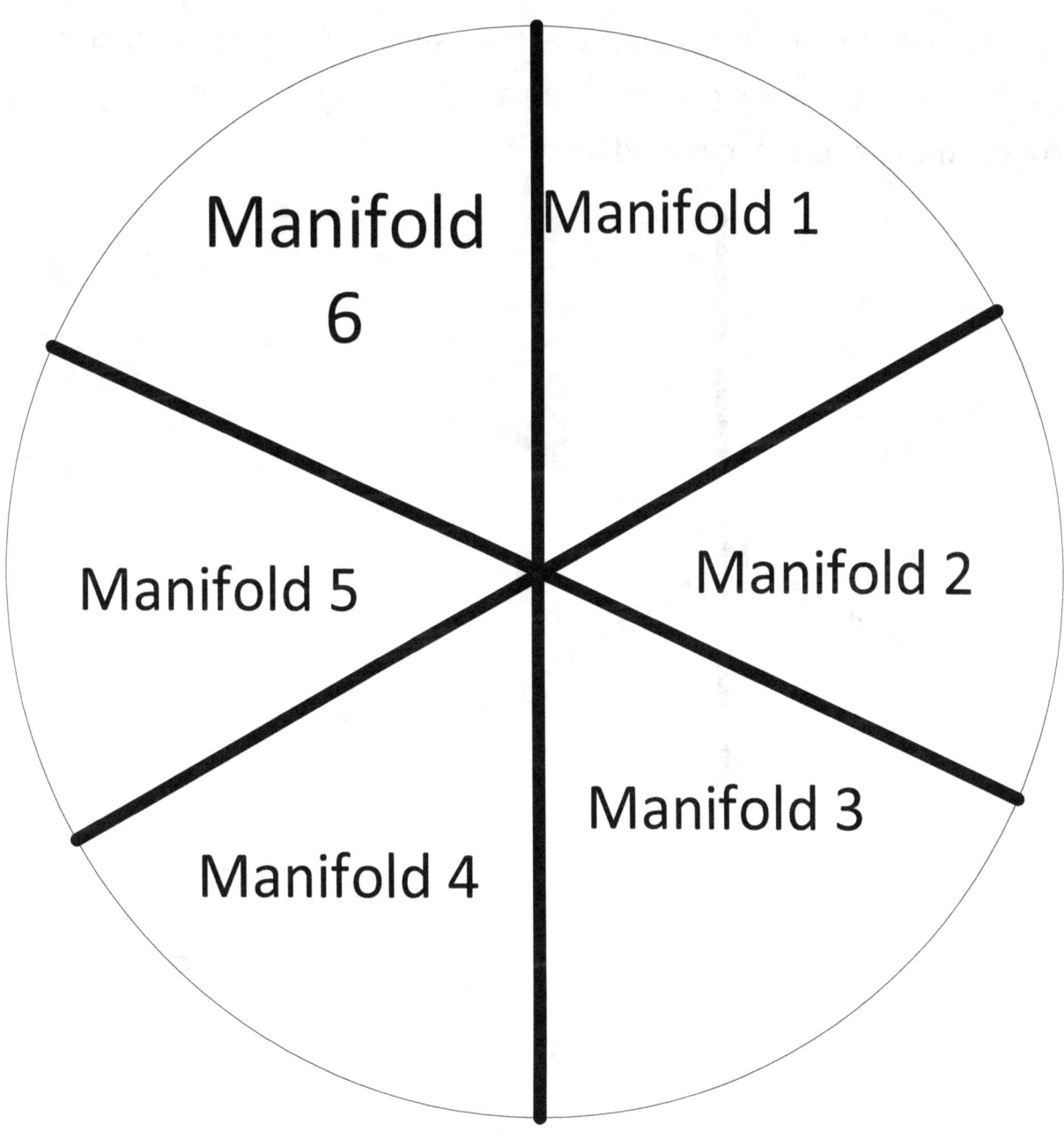

In this picture, manifold 1 is adjacent to manifold 6 and 2 . Manifold 2 is adjacent to 3 and 1. Manifold 3 is adjacent to 4 and 2. Manifold 4 is adjacent to 5 and 3. Manifold 5 is adjacent to manifold 6 and 4. And finally manifold 6 is adjacent to manifold 5 and 1.

A more robust thought experiment is to view the Big Bang singularity as a tiny sphere in its embryonic stage and exploding outward into a sphere. In this case the cleaving creates a series of concentric circles stacked on top of each other to form a sphere. Two such rotated concentric circles are pictured below:

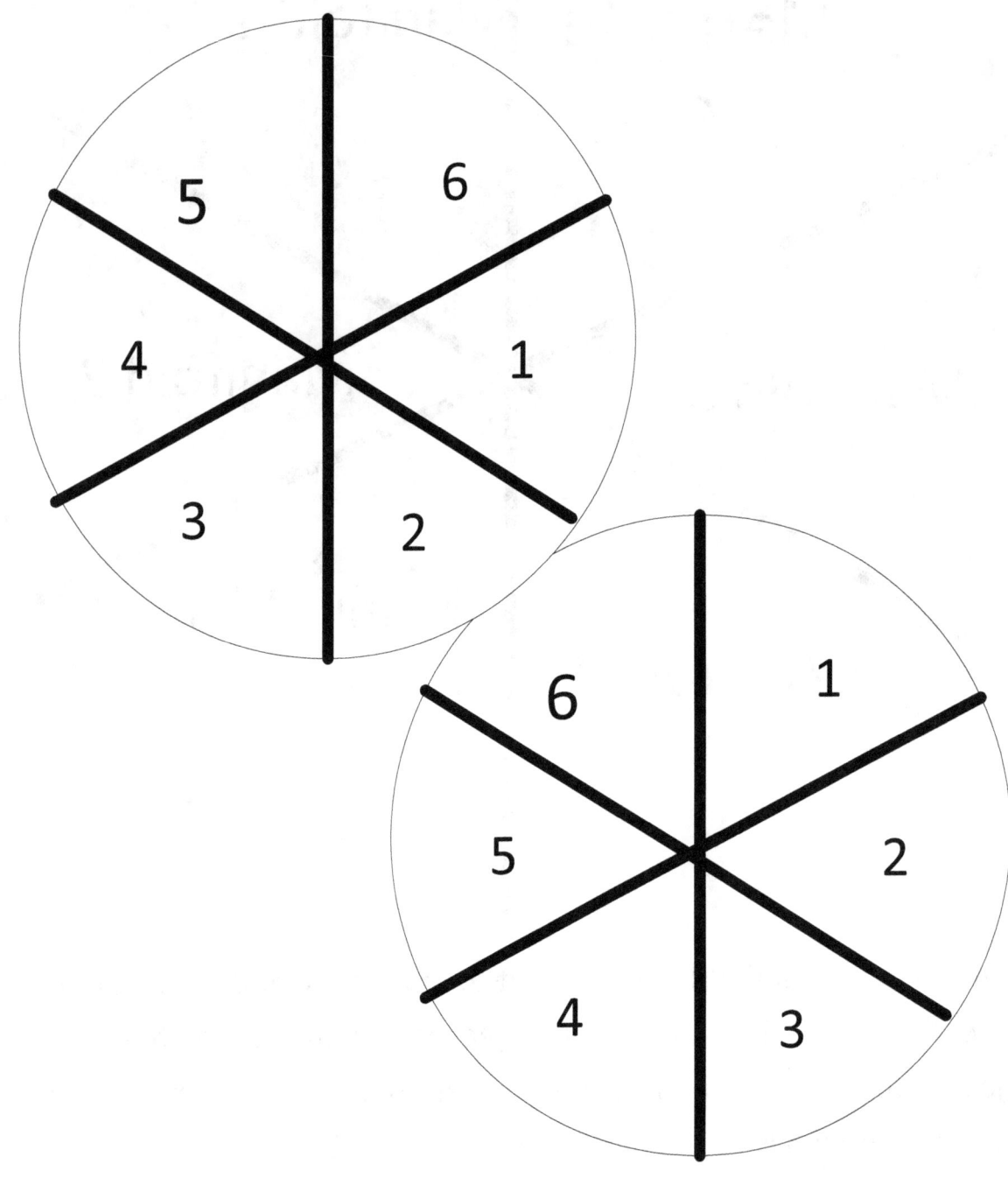

In this visualization, think of the same clock, but for each stack it gets rotated by 2 hours. So, the first clock starts at noon, the second clock starts at 2 pm, the third clock starts at 4 pm, etc. These clocks get stacked on top of each other (and below the central plane) to complete the sphere. Each concentric slice of the pie has an intermanifold energy layer not just radially as in the simplified case before, but also with the slice above it.

In this more complex example, manifold 1 is always adjacent to manifold 2 separated by an intermanifold energy in two perpendicular planar directions. The immediate neighbor model looks like 6/1/2, 1/2/3, 2/3/4, 3/4/5, 4/5/6 and 5/6/1. This visualization provides keen insight into how it is possible for quantum folds to provide a transition path like 1->2->3->4->5->6->5->4->3->2->1.

The notion of the 24-dimensional singularity exploding into a spherically symmetric layout with adjacent manifolds and separated by intermanifold energy is a very intuitive one and provides a clear roadmap on transition states.

Note how the arrow of time works in my model. In the simple view of each manifold being a sextant of the circle, think of the arrow of forward movement of time as pointing from the origin radially outward to the outer perimeter. This is true in each sextant.

In the more complex model, the arrow points from the origin of the sphere towards the outer edge of a section a region that has a number of these pie slices of the circle stacked together, true for each of the "pie slices".

In either visualization, if you do a thought experiment with time running backwards, you end up with a compression of the distinct spacetime manifolds into the singularity with 24 total dimensions at the time of the Big Bang. This is the conceptual view of the cleaving of the manifolds in reverse.

Let us see if my model :

1. Fits in with existing theories. It must collapse to the usual models in certain limits to have a chance at validity.
2. Measures up against the mysteries we explored in this book thus far. It will become apparent it provides conceptual consistency for many of the fundamental puzzles. The next sections will outline how this model provides some simple and reasonable explanations.
3. Lines up with existing observational data and suggests an experiment that can verify the accuracy of the model.

A. Mass Distribution in our Universe

According to current models, visible matter makes up about 5% of the universe. In our model, each spacetime manifold would consist of 5% visible matter, adding up to total of 25% for extra manifold matter. My thesis is that "dark matter" is just the extra manifold matter impacting us but attenuated by the intermanifold energy.

The remaining 70% of the universe is made of the energy between the manifolds. This is what in current models is referred to as "dark energy". This is remaining from the Big Bang but trapped in inter-manifold layers. It is distributed between manifolds 6-1-2 as follows:

MANIFOLD 6
5% visible mass local to manifold

INTERMANIFOLD

ENERGY

11.67 PERCENT

ENERGY OF UNIVERSE

MANIFOLD 1
5% visible mass local to manifold

INTERMANIFOLD

ENERGY

11.67 PERCENT

ENERGY OF UNIVERSE

MANIFOLD 2
5% visible mass local to manifold

This picture covers the 2 adjacent manifolds to manifold 1. Between the manifolds there is a layer of intermanifold energy left over from the Big Bang. The quantum folds and their interactions are shown via red lines and small tears in the fabric. We will go into the detail of the nature of quantum fold and the intermanifold energy in subsequent sections. This pattern repeats for all the manifolds. The result is:

- ➤ Energy layer between 1-2: 11.67%
- ➤ 5% local visible matter in manifold 1
- ➤ Energy layer between 2-3: 11.67%
- ➤ 5% local visible matter in manifold 2
- ➤ Energy layer between 3-4: 11.67%
- ➤ 5% local visible matter in manifold 3
- ➤ Energy layer between 4-5: 11.67%
- ➤ 5% local visible matter in manifold 4
- ➤ Energy layer between 5-6: 11.67%
- ➤ 5% local visible matter in manifold 5
- ➤ Energy layer between 6-1: 11.67%
- ➤ 5% local visible matter in manifold 6

The totals conform precisely to our observations of the universe: 5% visible mass, 25% dark matter (which is just the matter visible locally to each of the other five spacetime manifolds), and 70% dark energy (11.67% times 6 layers of intermanifold energy).

Note the beautiful symmetry in the universe that is maintained from the time of the Big Bang moving forward. To the extent that symmetry is broken, it is the decomposition of the pure energy composite universe(composed of six manifolds) into 3 matter dominated manifolds and 3 antimatter dominated manifolds after the explosion. The overall symmetry is kind of pretty and is maintained.

These energy layers are like hydrodynamic fluids or plasmas with all the complexity of systems with flows. Their eddies, undertows and buoyant forces are responsible for the microscopic movement through quantum folds (defined as a small piece of space time cycling through the manifolds 1->2->3->4->5->6->5->4->3->2->1)

B. Matter Antimatter Imbalance

This slight asymmetry explains our existence. Without it, we would be awash in a sea of photons and pure energy. Matter and antimatter annihilate each other and produce energy. The fact that our spacetime manifold has more matter than antimatter makes our existence possible. However, there is no good explanation as to why this is the case. For example, we could have equally well ended up with slightly more antimatter than matter- that would have still resulted in a working universe.

This is easily explained in my model by the following plausible scenario: the imbalances balanced each other out over the six-spacetime manifolds. After the Big Bang, manifold one, three and five experienced a slightly greater matter concentration, while two, four, and six experienced a slightly greater antimatter concentration. This is depicted in the following diagram:

MANIFOLD 6 Antimatter Dominant

5% visible mass local to manifold

INTERMANIFOLD ENERGY 11.67 PERCENT ENERGY OF UNIVERSE

MANIFOLD 1 Matter Dominant

5% visible mass local to manifold

INTERMANIFOLD ENERGY 11.67 PERCENT ENERGY OF UNIVERSE

MANIFOLD 2 Antimatter Dominant

5% visible mass local to manifold

INTERMANIFOLD ENERGY 11.67 PERCENT ENERGY OF UNIVERSE

MANIFOLD 3 Matter Dominant

5% visible mass local to manifold

INTERMANIFOLD ENERGY 11.67 PERCENT ENERGY OF UNIVERSE

MANIFOLD 4 Antimatter Dominant

5% visible mass local to manifold

INTERMANIFOLD ENERGY 11.67 PERCENT ENERGY OF UNIVERSE

MANIFOLD 5 Matter Dominant

5% visible mass local to manifold

INTERMANIFOLD ENERGY 11.67 PERCENT ENERGY OF UNIVERSE

MANIFOLD 6 Matter Dominant
5% visible mass local to manifold

Overall, the total would be completely balanced, which makes more sense for a compact form of concentrated energy as in the original primordial state at the time of the Big Bang. If you lived in manifold two, four or six antimatter would be the dominant form and matter would be extremely rare and produced in labs by theoretical physicists. The following chart shows a hypothetical split for illustrative purposes.

Manifold	%Matter	%Antimatter
1	99.999	.001
2	.001	99.999
3	99.997	.003
4	.003	99.997
5	99.998	.002
6	.002	99.998
Total (divide sum by 6)	50	50

The pool of energy between the manifolds prevents the manifolds from annihilating each other. Antiparticles or particles trying to breach the pool annihilate very quickly with the existing particles in the pool and just contribute to the energy pool.

The origin of this split is that right after the Big Bang, the manifolds started to crack apart. During this process, the quantum folds start to appear connecting all six manifolds. Randomly, antiparticles drift towards the quantum folds of manifold 2 , for example. Once that happens there will be a cascading effect. Any matter trying to drift in would annihilate with the antimatter at the fold and create energy. This would contribute photons to the intermanifold energy layer and inhibit matter from coming in.

It would also be a gateway for antimatter to flood in, since there is nothing inhibiting that. By random chance, ½ the manifolds would have matter as initial seeds and inhibit antimatter from entering while the other half would have antimatter as initial seeds and inhibit matter from coming in. Therefore, there is asymmetry in each manifold while maintaining a symmetry in the total number.

Incidentally, this removes one of the objections to the characterization of dark matter as ordinary matter. Namely, that if it were ordinary matter, the early light elements would have been influenced by them and cascaded into greater numbers than we observe. Since the fragmentation of the manifolds and the distribution of the initial seeds happens very quickly after the early moments in the Big Bang, there would be no cascade effect. Instead, there would be a uniform distribution of ordinary matter (or antimatter) across all the manifolds.

C. Quantum mechanics

The separation between the six spacetime manifolds is filled with intermanifold energy. We must be able to capture the effect of this separation in a quantifiable manner.

Thinking of the inter-manifold energy layer as a fluid, it has ebbs, flows and densities that vary how much separation it induces between manifolds locally. We need to quantify this intermanifold separation. This is accommodated by defining a scalar intermanifold separation factor as follows:

$$imf_{ij} = d_{ij} \{ [1 + (1/ d_{ij}^2)]^{1/2} - 1 \} \qquad\qquad [8.8]$$

where d_{ij} is a scalar factor that varies as a function of location in our spacetime manifold. For example, if you are in the far reaches of deep space but there is a large mass like a star in manifold 2 near you, the factor d_{12} will be small and imf_{12} will be near 1 (assuming our native manifold is manifold 1).

When d_{ij} tends to zero, the intermanifold factor tends to 1 as a limit. When d_{ij} tends to infinity, the intermanifold factor tends to zero and makes no contribution. imf_{ij} is similar to the familiar Lorenz contraction factor for particles with rest mass greater than zero in that it is bound between 0 and 1 in limits, but never actually achieves those values. As a concrete example, a value of 100 for d_{ij}, which is not particularly large, already reduces the factor imf_{ij} to .005, making the contributions for other manifolds approximately 1/200th the contributions of our own manifold. Let us say, on the other hand that the manifolds are awfully close and d_{ij} is .01, which is not particularly small, then imf_{ij} becomes .99.

A value of .99 results in pronounced intermanifold effects on the flat space metric, one that nearly matches our native spacetime manifold flat space metric. The general guidance for values of d_{ij} is that an extreme mass will distort the native manifold to such an extent that it will result in a high value for *imf*, while sparsely distributed low masses will result in an incredibly low value for *imf*. At quantum scales, whatever the value of *imf* there will be huge impact on the behavior of particles.

If this behavior of the intermanifold factor seems strange, imagine each spacetime manifold as a series of four-foot cement slabs with one-inch gaps between the slabs- this is something you will see walking on a pavement, for example. Then imagine the slabs stacked together with some water between them, with the cracks not lined up.

Now if you are a person, you will walk on top of the pavement and the gaps make absolutely no difference to your motion. Since your foot and shoe are substantially larger than the gap, your motion is the same as if the pavement were one continuous cement slab without cracks. This corresponds to the case where you live in your native manifold, and the intermanifold factor is negligible.

Now imagine you are an ant. Then the gaps look like massive chasms to you. To get from one point to another, you must navigate the gaps, where some undertow pulls you to another slab and you must walk with impaired motion due to water (our ant can survive under water). You keep dropping until you are at the lowest slab because you keep hitting undertows and gaps until you hit the lowest slab. Then some buoyant forces move you up slab by slab until you return to the first slab. The overall energy expended within the layers between the slabs would be net zero, because the forces pushing you up on the way up are nullified by the forces pulling you down. This would be normal motion to you.

To a person looking at your movement in the first slab, it will look like you disappeared and appeared at a remote point. In other words, it is hard to localize your position at any time. This scenario corresponds to the quantum particle case and the intermanifold factor is significant-the ant needs to traverse as far on the cement slabs as it needs to get to the next gap.

As peculiar as this movement might appear to you, it is in fact exactly what happens at the quantum level. Particles appear at one spot and another distinct spot without any evidence they travelled in between the spots.

Now imagine you are a massive truck. The slab bends downwards in the water, making your entry into the second slab easier. This corresponds to the intermanifold factor becoming larger again for massive objects. Admittedly, this is not a perfect analogy, but it gives you an idea of how the intermanifold scaling works in a 24-dimensional universe.

Factoring in intermanifold effects from the other manifolds, we can write the generalized displacement in our spacetime manifold as:

$$ds_{11}^2 = ds_1^2 + (imf_{12}ds_2)^2 + (imf_{13}ds_3)^2 + (imf_{14}ds_4)^2 + (imf_{15}ds_5)^2 + (imf_{16}ds_6)^2 \qquad [8.9]$$

The concept of shadow displacement is important here. The units of distance in manifold 2 will be different, but the apparent displacement in our manifold due to the contribution from manifold 2 is equivalent to the second term in this equation. Here is a diagram for sample displacements involving the first two manifolds.

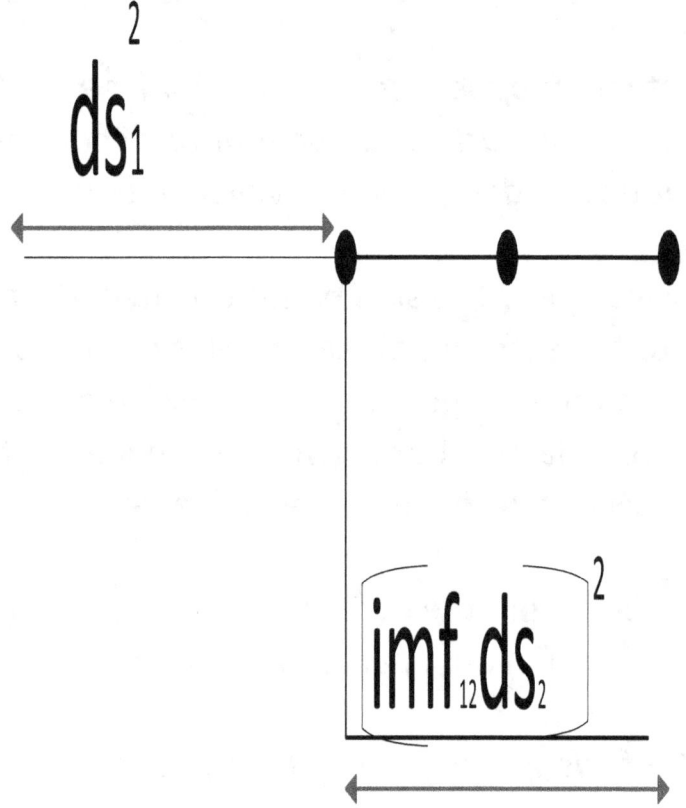

$$ds_1^2$$

$$imf_{12}.ds_2^2$$

FIGURE 1 Displacement in an adjacent manifold

So, the displacement in manifold two which will differ from light seconds has an equivalent shadow displacement in light seconds, represented by the dotted line. The movement in detail would be a transition from manifold 1-2, followed by a displacement in 2 , and cyclically repeating 1->2->3->4->5->6->5->4->3->2->1 . When the particle returns to manifold 1, the apparent displacement is just a sum in light seconds of the quantities in this equation. This is the basic movement through a quantum fold in my model. The following diagram shows the fundamentals of displacement through a 24-dimensional universe.

QUANTUM FOLD

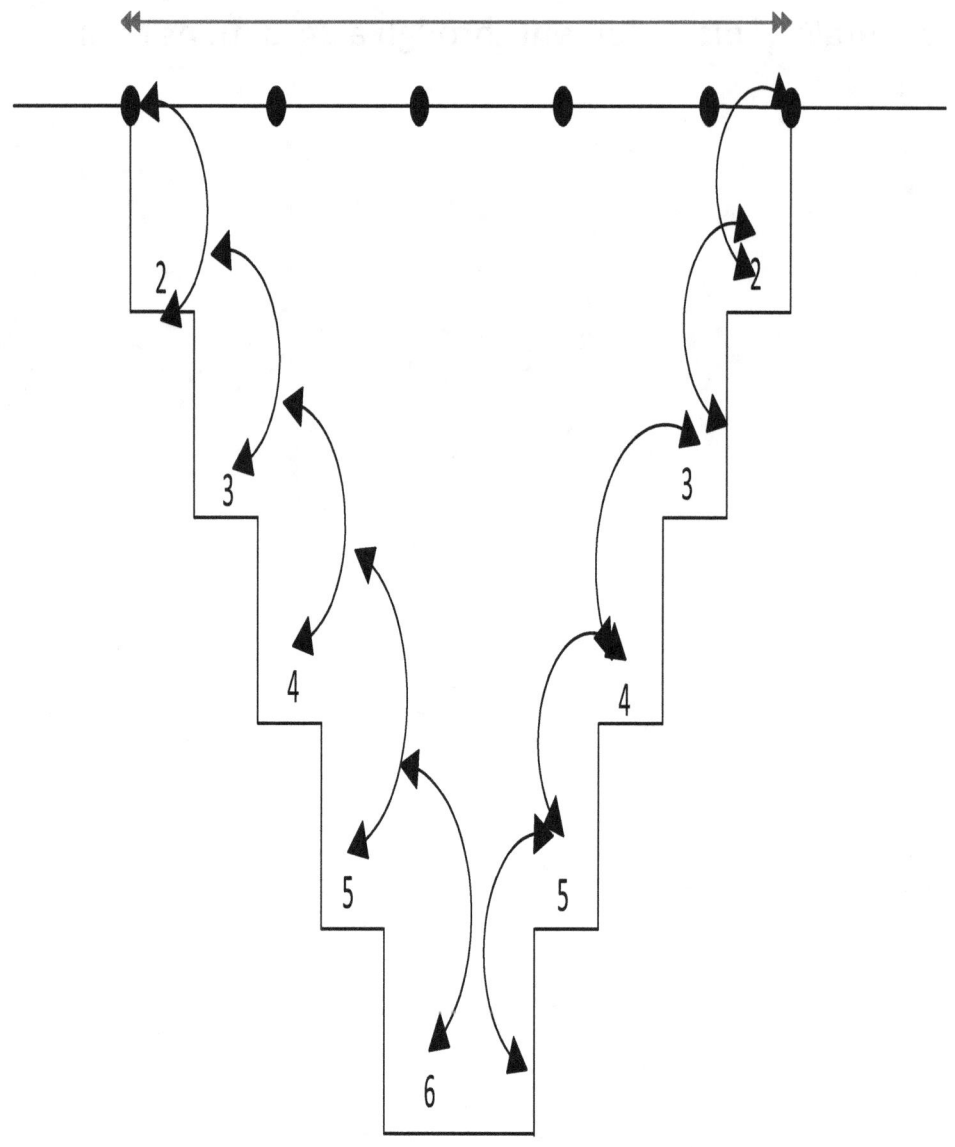

FIGURE 2 A quantum fold, the fundamental building block of the univerese

The level step is movement through manifold 1. Then there is an abrupt drop, which is a transition through an intermanifold energy layer to manifold 2. The energy layer is like a hydrodynamic fluid and the abrupt pull is akin to an undertow. Once at the second step the displacement is $(imf_{12}ds_2)^2$. This pattern repeats until you drop down to manifold 6 where the displacement is $(imf_{16}ds_6)^2$. This cycle is followed by a buoyant push up manifold by manifold until you get back to the level manifold 1. The entire apparent movement in manifold 1 is represented by the dotted line. That defines the quantum fold of the universe- the very tiny hole through which quantum particles slip in and out of manifolds only to ultimately return to their 'native' manifold.

Before you dismiss this motion as nonsensical, remember that quantum motion is very much like this. The classic case of an electron "tunneling" through a potential barrier, is a case of an electron being on one side of the barrier, not in any way climbing the barrier and then appearing on the other side of the barrier. Think of the barrier as a hill. The electron sits on one side of the hill , never climbs the hill , provides no evidence it tunneled through the hill and magically appears on the other side of the hill. My description of shadow displacement and the dotted line would be the dotted line connecting one side of the hill to the other.

Figure 2 shows the fundamental nature of displacement at quantum levels. The dotted line represents the shadow displacement in our manifold. For the entire duration of the quantum fold (the length of the dotted line) , the particle is not in our manifold.

The probability of the particle being in the interval $(imf_{12}ds_2)^2$ is going to be :

Probability of being in manifold 2= ($imf_{12}ds_2$)² /(($imf_{12}ds_2$)²
+($imf_{13}ds_3$)² +($imf_{14}ds_4$)² +($imf_{15}ds_5$)² + ($imf_{16}ds_6$)²)   *[8.10]*

This is simply the displacement squared in manifold 2 divided by the total displacement squared across the quantum fold. There are similar equations for all the other intervals. Note that my model provides a physical foundation for why motion at quantum scales appears inherently probabilistic. For example, an electron jumping from one quantum state to another without traversing in between is like going from START to FINISH across the quantum fold.

The very nature of traversing through the manifolds and returning makes the entire journey appear to be a probabilistic one in a single manifold. Our explanations in quantum mechanics arise from the fact that although endpoints of a quantum fold correspond to physical observables, all the states in between are in transit and are immeasurable.

Quantum mechanics uses wavefunctions and has them collapse into eigenstates to explain nature, without any real explanation of why it works.

The framework in my model explains why that sort of formulation works. Eigenstates are real measurement points in our manifold separated by the gulf of a quantum fold or multiple folds. Between eigenstates are quantum folds that inhibit precise measurement. The quantum folds make it difficult to measure any properties within the

The level step is movement through manifold 1. Then there is an abrupt drop, which is a transition through an intermanifold energy layer to manifold 2. The energy layer is like a hydrodynamic fluid and the abrupt pull is akin to an undertow. Once at the second step the displacement is $(imf_{12}ds_2)^2$. This pattern repeats until you drop down to manifold 6 where the displacement is $(imf_{16}ds_6)^2$. This cycle is followed by a buoyant push up manifold by manifold until you get back to the level manifold 1. The entire apparent movement in manifold 1 is represented by the dotted line. That defines the quantum fold of the universe- the very tiny hole through which quantum particles slip in and out of manifolds only to ultimately return to their 'native' manifold.

Before you dismiss this motion as nonsensical, remember that quantum motion is very much like this. The classic case of an electron "tunneling" through a potential barrier, is a case of an electron being on one side of the barrier, not in any way climbing the barrier and then appearing on the other side of the barrier. Think of the barrier as a hill. The electron sits on one side of the hill , never climbs the hill , provides no evidence it tunneled through the hill and magically appears on the other side of the hill. My description of shadow displacement and the dotted line would be the dotted line connecting one side of the hill to the other.

Figure 2 shows the fundamental nature of displacement at quantum levels. The dotted line represents the shadow displacement in our manifold. For the entire duration of the quantum fold (the length of the dotted line) , the particle is not in our manifold.

The probability of the particle being in the interval $(imf_{12}ds_2)^2$ is going to be :

Probability of being in manifold 2= $(imf_{12}ds_2)^2 /((imf_{12}ds_2)^2$
$+(imf_{13}ds_3)^2 +(imf_{14}ds_4)^2 +(imf_{15}ds_5)^2 + (imf_{16}ds_6)^2)$ *[8.10]*

This is simply the displacement squared in manifold 2 divided by the total displacement squared across the quantum fold. There are similar equations for all the other intervals. Note that my model provides a physical foundation for why motion at quantum scales appears inherently probabilistic. For example, an electron jumping from one quantum state to another without traversing in between is like going from START to FINISH across the quantum fold.

The very nature of traversing through the manifolds and returning makes the entire journey appear to be a probabilistic one in a single manifold. Our explanations in quantum mechanics arise from the fact that although endpoints of a quantum fold correspond to physical observables, all the states in between are in transit and are immeasurable.

Quantum mechanics uses wavefunctions and has them collapse into eigenstates to explain nature, without any real explanation of why it works.

The framework in my model explains why that sort of formulation works. Eigenstates are real measurement points in our manifold separated by the gulf of a quantum fold or multiple folds. Between eigenstates are quantum folds that inhibit precise measurement. The quantum folds make it difficult to measure any properties within the fold, so we must force a "fuzzy look" interpretation on it and make it collapse into observable states at endpoints of the quantum folds.

Regardless of who/what/why the measurement itself is being performed, the real physical observables are the same. This addresses numerous philosophical problems in terms of measurement associated with quantum mechanics.

To determine the path of an electron, quantum mechanics envisions describing it with wavefunctions (that cover all of space !!), probability coefficients and collapse of wavefunctions into observable states.

In my model, the same kind of probabilities can be constructed without all the mysterious math with no physical meaning. As a particle transitions from manifold 1->2->3->4->5->6->5->4->3->2->1 , its quantum states can be undergoing transformations within each extramanifold before it arrives back in its native manifold in a different state.

Note carefully that a probabilistic view of the motion of a particle emerges very naturally here. The probability of a particle being in a manifold depends upon some coefficients that indicate that it was temporarily traversing in an alternate manifold in a quantum fold. It eventually pops back into the "native" spacetime manifold.

When you zoom in on the fine grain nature of the universe, it automatically becomes a probabilistic event to even locate the particle! This is exactly in line with our quantum mechanical view of the universe.

You can keep the entirety of the mathematical framework of quantum mechanics- or better yet, craft a system of probabilities based on this model that will ultimately turn out to be a more elegant formulation.

Thus, the seemingly mysterious probabilistic view of particle states that quantum mechanics suggests is just a natural outgrowth of the 24-dimensional model.

D. The Measurement Problem and the Uncertainty Principle

There are various ways that the measurement problem is stated, but the core idea is that the observed physical state of the system is determined by a collapsing wavefunction. The wavefunction itself is a superposition of several states in an extremely complicated mathematical space with probability coefficients.

This has led to several questions about interpretation. What constitutes a measurement? Is a human being necessary for the observation process?

Viewed in the following way, I think we can eliminate the interpretation problem: Any quantum mechanical particle that we think is existing solely in our spacetime manifold is transitioning between manifold 1, 2, 3 4, 5 and 6 continually. Because we only glimpse a sliver of this reality, we perceive a measurement problem.

 If you could somehow view the path of the particle across all the spacetime manifolds, there would be no need to worry about when/how/where it collapsed. It would appear as more of a smoother path with no need to worry about measuring to collapse it. We need to recast the math of quantum mechanics to reflect this type of behavior – and we should be able to retain all the current results with no difficulty.

To gain more insight into the measurement problem and how my model eliminates it, consider a canonical quantum mechanical problem- the quantum state of an electron, which you can describe with four quantum numbers- principal, angular, magnetic and spin. The figure below shows how an initial and final state of an electron are connected.

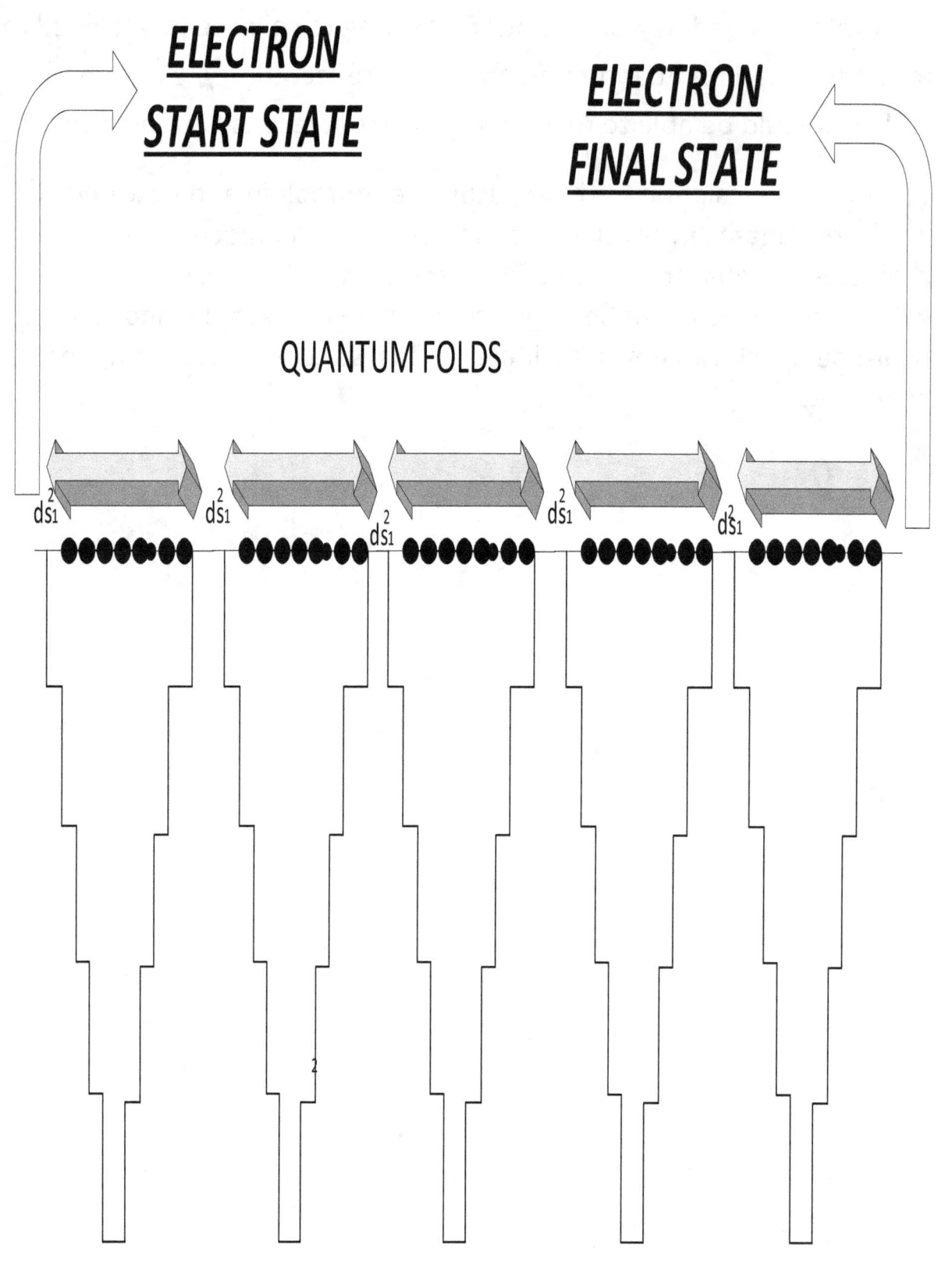

The standard measurement paradigm in quantum mechanics views the electron as a wavefunction that is a superposition of all possible states with probability coefficients.

In this model, the electron collapses into its final state when we measure it- this is the physical world measurement. As strange as this sounds, it yields fantastically accurate answers.

This is what causes most physicists to shrug their shoulders and accept the model.

However, in my model, we can end up with the same physical result without having such extreme superposition principles. The electron starts in a state, then it drops through a quantum fold. In this process it interacts with the energy layer.

This cycle repeats 1->2->3->4->5->6->5->4->3->2->1 , possibly multiple times, with a negligible displacement in the native manifold, and finally lands in manifold 1 in its final state. The measurement is a real one. The electron is real, the energy it gains is real via the intermanifold energy layer. The final state is real. It is still possible to have different final states because that is a dependency on what happens in each transition cycle and the interactions with the energy layer.

 We cannot measure in its transition states because it is doing its cyclical dance through the manifolds. This model still contains the uncertainties in the transition cycles and there is still the need to factor in probabilities at a very fundamental level. The need for superposition of all states in traditional quantum mechanics makes sense in this picture because you must factor in all possible interactions along the way to determine the variable final state. This has all the same properties as predicted by quantum mechanics but is a fundamentally different model.

This view also coincides with the quantum field theoretic proposition of the vacuum having energy. Except what is labelled as vacuum energy is the intermanifold energy.

Now we turn to the uncertainty principle. This principle has two main forms. The first one considers energy and time as conjugate pairs that cannot be simultaneously determined with great accuracy. The second one considers position and momentum as a conjugate pair and asserts similar conclusions.

Ironically, the uncertainty principle has a clear prescriptive formula associated with it. This itself raises the question of why the universe is so precisely imprecise, if you will pardon the turn of phrase.

If the universe were truly uncertain, one might expect some deviation in a rule of this form.

Let us look at the case of one dimension in space and an associated momentum in that direction. The uncertainty principle states that:

$$\Delta x \, \Delta p_x \geq \hbar/2 \qquad\qquad [8.11]$$

In this equation $\hbar$ is Planck's constant **h** divided by 2π. The value of the Planck's constant is **6.62×10^{-34}** J-s. The more precisely we figure out the position, the smaller Δx becomes and the larger Δp_x has to be to compensate for it.

And conversely, the more precisely we determine the momentum, the larger the uncertainty in the position. The universe always forces the combined precision to be greater than a certain fundamental constant.

Let me completely repaint this picture in my model while explaining exactly why the uncertainty principle works the way it does. The universe is saying there is small value where everything is uncertain (less than the order of Planck's constant).

The existence of such a minute gap of complete uncertainty fits precisely with my model. For any given quantum object, the momentum has extents into all six spacetime manifolds, as does the position.

This is indicated in our coordinate system- both are components of a 24-dimensional space. The generalized momentum will be a 24-component tensor- in a single spacetime manifold it is a 4-component tensor.

This geometry leaves a gap in the form of the quantum fold. Anything trapped in the gap is out of scope for any type of accurate measurement. Another inference is that there is a fine grain limit on the precision of measurement- right in line with the uncertainty principle.

To get an idea of how the uncertainty principle fits in my model, we are going to invert the problem and determine for what chunks of displacement will the position and momentum be completely uncertain. Note that we can write the uncertainty principle as :

$$\hbar/2 < \Delta x \, \Delta p_x \qquad\qquad [8.11A\,]$$

That will determine the "box" that uncertainty must lie outside of. I am looking at the uncertainty principle in a complementary manner. This tells us the size of the box $\Delta x \, \Delta p_x$ must be in terms of a hole in spacetime. If we suppress 2 dimensions and set the quantum fold displacement as $\{imf_{12}ds_2)^2 + (imf_{13}ds_3)^2 + (imf_{14}ds_4)^2 + (imf_{15}ds_5)^2 + (imf_{16}ds_6)^2\}$ $= c^2 \Delta t^2 - \Delta x^2$ and we have the first term set to magnitude of $\hbar^2 + \epsilon$ and the second term is set to magnitude $\hbar^2$ divided by 2 we see that the size of our "hole in space" is the magnitude of $\hbar^2$ divided by 2 plus ϵ . ϵ is an arbitrary exceedingly small number here to enforce the inequality in equation 8.11A and will be ignored henceforth.

With this constraint, we see that if **Δx** lies inside the "box of uncertainty" then so does **Δp$_x$** . On the other hand, if either is outside the box, you are out of the quantum fold and you can start to get definite values. This aligns correctly with the uncertainty principle.

Our quantum particle moving in the x direction has an extent in manifold 2, 3, 4, 5 and 6. Recall that even in flat spacetime, there is a component of the metric extending into the other manifolds.

For exceedingly small changes, these factors become prominent. If it starts at position x and moves to position x+ Δx and we can tell what it is, it means that it has moved past the quantum fold. On the other hand, if it falls anywhere within the quantum fold, the position becomes completely unknown and immeasurable.

If you want to look at it mathematically, the x component of the momentum in the first coordinate position shifts to the fifth coordinate position. This dance of the quantum object cycles through all the manifolds until it returns to manifold 1.

The same logic applies to the momentum. Since it depends on position with respect to time, it will have no meaning when the time component of the spacetime manifold shifts. So, inside the quantum fold gap it will become completely uncertain, but outside you should hit the realm of measurability.

Mathematically, as you shift to the second set of manifold coordinates you can no longer evaluate the coordinate derivatives using the first set.

Consequently, I can draw an area of complete uncertainty as follows:

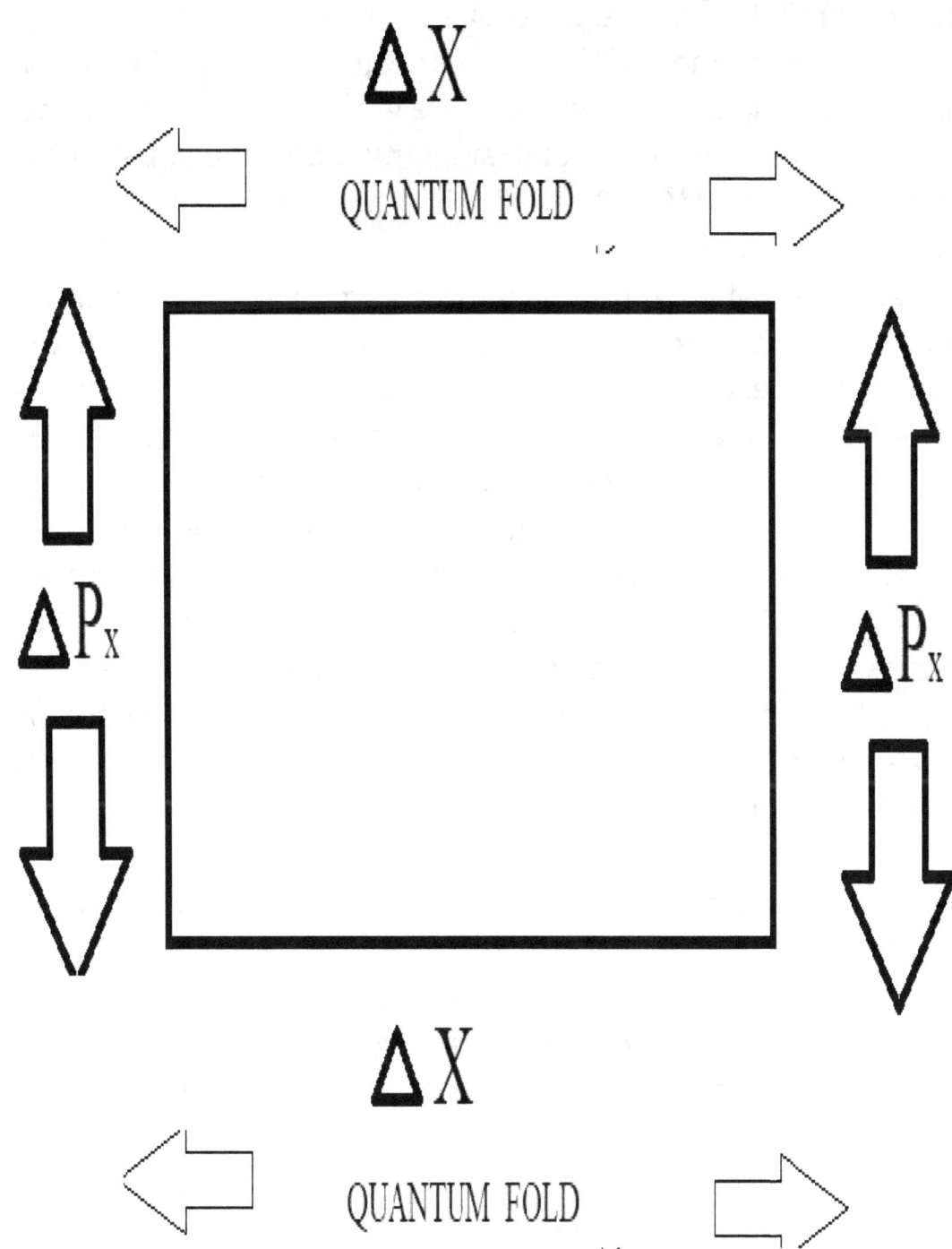

The fine grain nature of the universe suggests that the quantum fold is on the order of 10^{-17} in one dimension. This is a fundamental feature of the universe in my model. The size of the quantum fold does not vary except in extreme circumstances where gravity becomes overwhelming (like a black hole). Thus , for almost all situations, the applicable area of the uncertainty is 10^{-17} x 10^{-17} which is in the right order of magnitude for Planck's constant.

Note carefully that my model suggests the conjugate uncertainty automatically along with the position uncertainty. Since the momentum depends on time, and the time is in the first coordinate in a spacetime location, the conjugate variable of momentum becomes uncertain as well. This gives us the "square of uncertainty" (10^{-17} m) x (10^{-17} kg-m/s) ~10^{-34} J-s, giving us the uncertainty principle. In my model, I am postulating that the displacement of the quantum fold squared is the magnitude of $\hbar^2$ /2. The magnitude of the quantum fold squared coincides precisely with one of nature's fundamental constants. This in turn results in the uncertainty principle, as uncertainty in space and the conjugate uncertainty in momentum are trapped in this fold and the size of the fold enforces the uncertainty principle.

Note that this extent refers only to one dimension and does not consider the other 2 spatial dimensions or the time-as-a-distance dimension. If we generalize this to all dimensions, we conclude there is a hole in spacetime in our manifold that prevents us from observing fine grain properties of matter and leads to the uncertainty principle.

The inequality forces you to be outside this box, because inside the box you are inside manifold 2-6, while outside you can take meaningful measurements of physical quantities.

I am postulating that $\{(imf_{12}ds_2)^2 + (imf_{13}ds_3)^2 + (imf_{14}ds_4)^2 + (imf_{15}ds_5)^2 + (imf_{16}ds_6)^2\} =$ **magnitude of $(\hbar^2/2)$.** This implies the inequality $\Delta x <$ **magnitude of $(\hbar/\sqrt{2})m$** in a one-dimensional case and consequently the uncertainty principle. The fact that multiples of the quantum fold represent eigenstates and measurable quantities aligns with quantum mechanical views of measurables being multiples of Planck's constant.

The quantum folds are areas in spacetime that prohibit movement within them from the viewpoint of the resident manifold. Spacetime in our manifold is a cube in four dimensions like a giant crystal with spacings separating the lattice points. Only the vertices of the lattices correspond to points in our resident manifold. Between the vertices of the lattices are quantum gaps on the scale of Planck's constant(cumulatively) that extend into manifold 2-6. The vertices represent points that are physically present in our manifold. These are real observables, with no relationship to how they are measured.

Every manifold will have their own version of the uncertainty principle as this cycling goes on. If you could somehow get an omniscient view of the position as it tracks along every spacetime manifold, it would not look uncertain to you.
Viewed in this light, the uncertainty principle is just another statement that our view is confined to our local spacetime manifold. Let us look at the energy-time equation:

$\Delta E\, \Delta t \geqslant \hbar/2$ [8.12]

Recall that the four momentum in our manifold is: (p^0, p^1, p^2, p^3) where we have the first component $p^0 = E/c$ and the rest of the components are the momentum, relying on the time coordinate. Applying a similar logic as for position, we have $\Delta E \sim (10^{-17}$ J $)$ and $\Delta t \sim (10^{-17}$ s$)$. We must adjust for factors of c in both the momentum and energy definitions .

Consider energies and time inside the following box in a quantum fold:

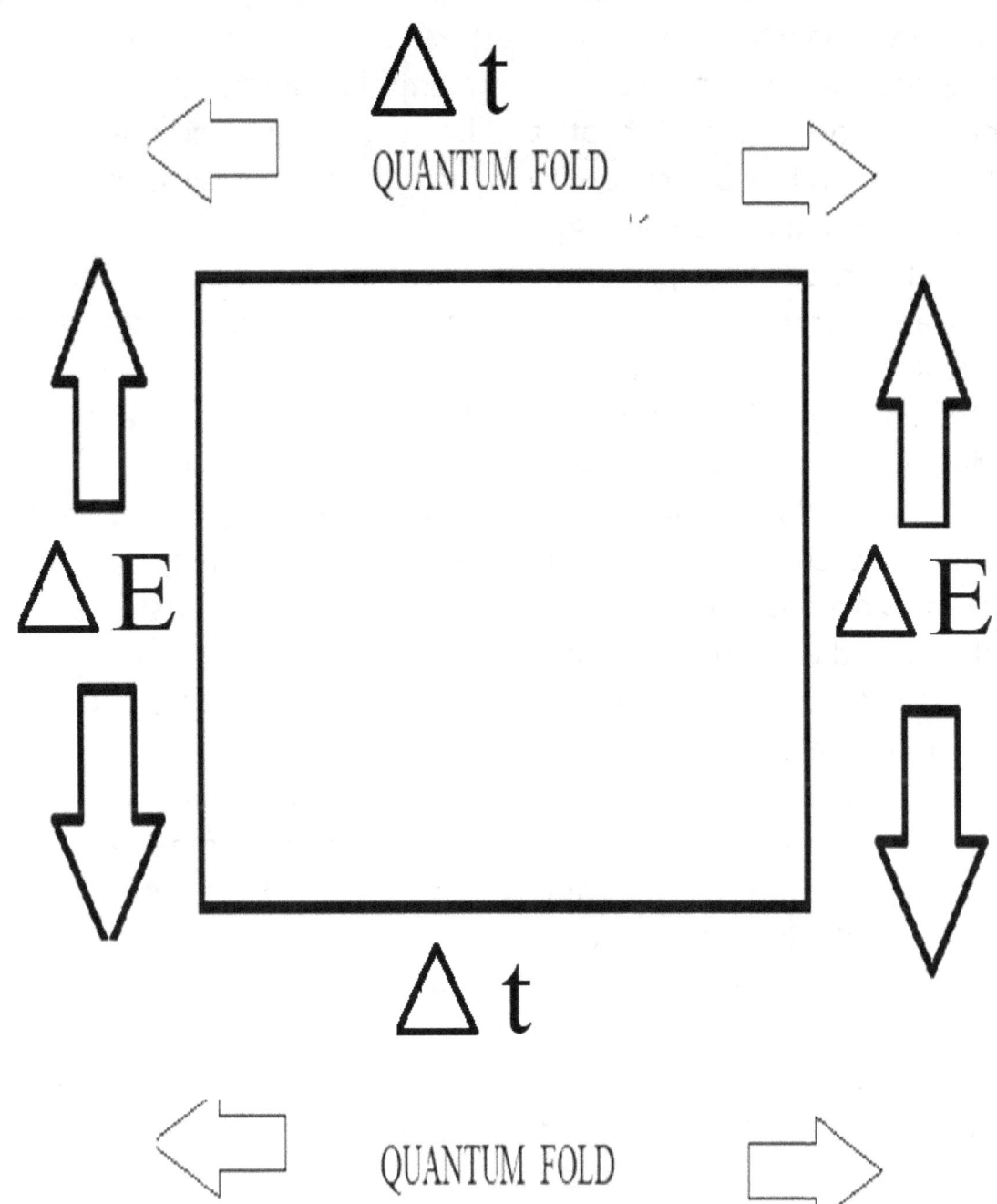

Inside the fold, the coordinates have shifted outside the first 4 coordinates; consequently, you cannot determine the slices of time anymore. In manifold 2 , for example, you are dealing with coordinate t_1 instead of t_0 as in manifold 1.

As for energy, being immersed in intermanifold energies during the journey, the particles can "cheat" and borrow energy. This is exactly what the uncertainty principle states: for small enough slices of time, you can cheat and borrow energy.

My model provides an explanation of where quantum particles are "temporarily" borrowing energy from- the intermanifold energy layer. The explanation that the uncertainty principle allows energy to be borrowed on credit for imperceptibly short intervals has a real physical foundation. It is real energy from intermanifold energy layers.

Note again that the conjugate uncertainties appear organically here- there is no treacherous logic involved in trying to get energy and time paired up. Since four momentum is tied to energy in the first coordinate and we know the variable ct is involved in the first coordinate of the position itself, it works out readily that both are uncertain together.

The uncertainty principle, in a nutshell, reflects our inability to measure physical quantities in other manifolds. There is nothing magical about these quantities, they are the same in each manifold. It is just that every manifold is in a cocoon separated by an intermanifold energy layer and clocks that march to the tune of different chronological drummers.

Every manifold has their own version of the uncertainty principle since they all suffer from the same handicap in terms of measurement of physical quantities across manifolds.

The key point here is that motion viewed in the totality of the 24-dimensional combined fabric suffers from no uncertainties. The uncertainties only manifest themselves when you are confined to a single four-dimensional spacetime manifold, as we are by necessity.

E. Quantum Tunneling

Quantum tunneling is another feature of quantum mechanics that is essential to the explanation of many well-established technologies. The existence of the Sun, which depends upon nuclear fusion, would not be possible without this basic feature.

It is also the foundation of tunnel diodes, quantum computing, and the functioning of transistors. It is a seemingly very bizarre phenomenon, where an electron can tunnel across a potential barrier that is seemingly impossible to climb classically.

Interpretations of this phenomenon rely on wave particle duality-another difficult concept to grasp. The idea that an electron can pass across a barrier without having to climb it is an odd one, as is the notion that it simply disappears from before the barrier and simply reappears after the barrier!

The standard quantum mechanical interpretation is that it has somehow tunneled through. This is reliant on the wave particle duality of the electron. If you view it as a little bowling ball trying to make its way up the hill, it lacks the energy to make it. If you look at it as a wave that has differing amplitudes on both sides with a probability of appearing on the other side, you can make the math work.

Let us look at how this can be viewed in my proposed model. Instead of having the electron doing gymnastics that do not make sense, it moves along from manifold to manifold as it makes its way across the potential barrier. Bear in mind that all motion is through six different four-dimensional spacetime manifolds.

As a simple example, to illustrate the point, consider a particle moving in one dimension trying to vault over a potential barrier. At time t in manifold one it is at position x to the left of the barrier. At t + Δt, assume it tunneled through and appears to the right of the barrier shaped like a hill. This is shown in the figure below.

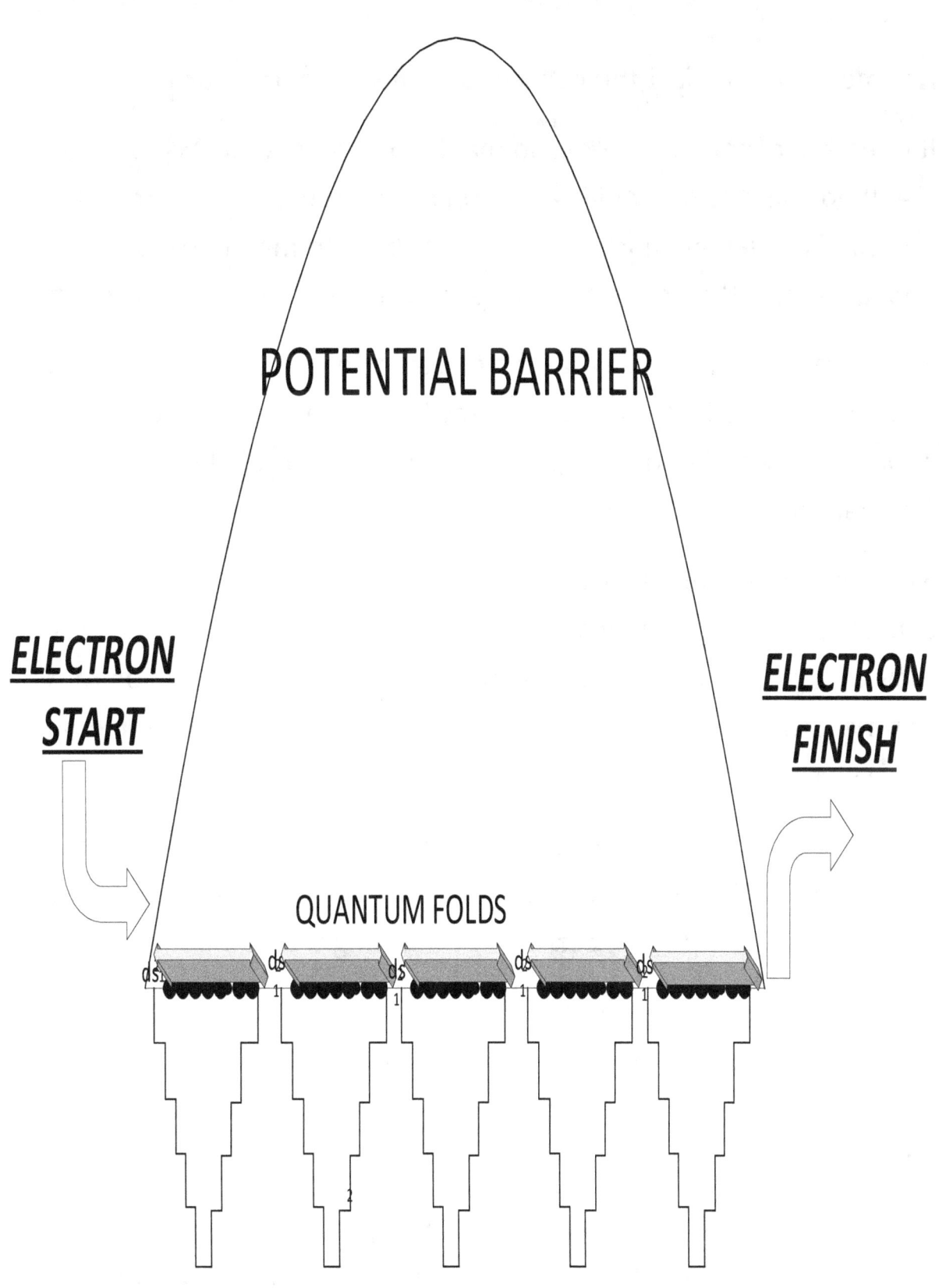

At time t in manifold 1 the particle is at the left of the barrier.

It does in fact appear at x+ Δx_2 in manifold 2, then at x+ Δx_2 + Δx_3 manifold 3 and so on until its position is far enough to be across the potential barrier, at which point it cycles back to manifold 1 and appears in just the right spot to appear as if it had tunneled through.

Note also that the energy that the electron needs to "tunnel" through is coming from the intermanifold energy layer. We can replace what appears to be an impossible process and give it a real physical foundation.

None of the mathematical solutions need to change, since this conforms to our observed actions – just our view of how the particle moves across multiple spacetime manifolds. The math could be recast to reflect this behavior and I suspect ultimately it will look simpler and more elegant when viewed in this light.

By way of summary: as you journey through subsets of 4 dimensional spacetime manifolds in our 24-dimensional universe, quantum particles can get across a potential barrier by traversing the quantum folds.

A good visualization for this is to pretend you are swimming on the top of the ocean surface, in calm waters and you see a point about a mile away. To get to it, you need to expend a lot of energy against the waves and tide.

An undertow pulls you down abruptly, you get swept up by the eddies and currents that are at the bottom of the ocean, and you get buffeted around. Some buoyant forces pop you up to the surface you saw when you were swimming in calm waters.

If you look only from the point of view of the surface of the ocean, it looks very much like you made the one-mile trip without expending any energy. It also appears like you disappeared at your origin and appeared at your destination.

In fact, a great deal of traveling was done, a lot of energy was expended in both pulling you down and pushing you up. If you did not know all the forces involved, you might be tempted to say you "tunneled" from your origin to your destination.

F. Quantum Entanglement

One of the most perplexing features of modern quantum theory is the idea of quantum entanglement. This is the notion that particles that are not spatially close can have their quantum properties like spin, polarization, position, and momentum correlated. It turns out that a photon here on Earth and a photon in a nearby star can be entangled so that even though they are many light years apart, their properties are 100% correlated. For example, if the photon here had spin up and it is entangled with a photon near Alpha Centauri, then the Alpha Centauri photon would also have spin up consistently 100% of the time (or spin down consistently 100% of the time).

The point is that despite the apparent total lack of connection between a photon here and a photon in Alpha Centauri, somehow their states are correlated. The more you think about this, the worse your headache might get. This mind-bending phenomenon has been experimentally verified for many particles at huge distances. One such experiment even involves using source photons from widely separated quasars billions of light years away from us!

When you examine or think about a phenomenon like this, even if you accept its experimental validity, it really challenges your imagination. How on earth do photons separated by huge distances "know" the state of the other one? It is inexplicable, and it has come to a point where no one tries to explain it- it is just accepted as reality. And it is the foundation of the embryonic field of quantum computing- the qubits in quantum computing are entangled, and that is the foundation of state knowledge of them.

In my model, I believe this can be explained in a clear way. To set the stage, think of a pipe with two ends. One end has a single valve that only controls the ratio of hot to cold water.

The other end has 2 valves, and you can turn on both or each individually. The overall pressure in the pipe is to remain constant. If you now open the cold-water valve at one end to its maximum, the other end with temporarily increase the hot water pressure and you get scalded if the water is running there.

This is like a situation where if the toilet flushes and you are in the shower, you notice a sudden burst of hot water (if you have a pressure sensing valve). This is not because the individual water molecule at one end somehow "knows" the state of the water molecule at the other end. It is because the pressure valve senses a reduction in cold-water pressure at one end and compensates.

Note that the very structure of my model guarantees that the intermanifold energy layer operationally acts as a nonlocal link between particles that may be widely spread out in their local manifolds. This is a consequence of the fact that each manifold consists of a series of quantum folds connected to each other like a giant crystal lattice. And underneath the quantum folds is the layer of intermanifold energy as a substrate across the entire manifold.

The following diagram shows you a single photon transitioning through a quantum fold between manifold 6 and manifold 1:

MANIFOLD 6
5% visible mass local to manifold

INTERMANIFOLD

ENERGY

11.67 PERCENT
ENERGY OF UNIVERSE

MANIFOLD 1
5% visible mass local to manifold

INTERMANIFOLD

ENERGY

11.67 PERCENT
ENERGY OF UNIVERSE

MANIFOLD 2
5% visible mass local to manifold

As a simplistic case, this throws the balance of spin off in both manifold 6 and manifold 1. Manifold 6 now has one net down spin and manifold 1 has one net spin up. The actual situation is more complex, because there are transitions between all manifolds more dynamically, but this will get the point across.

To compensate, manifold 1 must now send a spin down back to manifold 6. Or the intermanifold energy layer must supply another photon. This must happen as an event that is correlated, otherwise we are in an unbalanced situation. This simplistic example gives us an intuitive feel for how this balancing act makes it appear as if two completely unrelated photons can look like they have correlated states.

Returning to our pipe pressure analogy, in our universe, the pipe is the inter-manifold energy layer. At the time of the original Big Bang assume a distribution of x percent of photons in spin state up and y percentage in spin state down with the total of x and y being 100.

In a sensible universe, this overall global number between all manifolds that cleaved off should not change arbitrarily. That is the root cause behind quantum entanglement. The intermanifold layers are constantly providing a supply of photons across manifolds, which by the way, is completely consistent with quantum field theory assumptions.

The inter-manifold layers are like pressure regulators that must keep the overall properties in balance. It is not an issue of the photon on earth "knowing" the state of the photon at Alpha Centauri. That just seems ludicrous and irrational. It is more like the overall spin state across manifolds 1-6 is maintained in a neutral equilibrium.

So, the correlation is a result of a global conservation law that makes common sense, not some mystical knowledge. Let us say for example, both photons end in a spin up state. It might exactly balance out two photons there were in transition to manifold 2 where they end up spin down.

Here is a hypothetical simple distribution that will get the idea across. Assume x and y from the Big Bang were evenly distributed across all manifolds and the intermanifold energy layers. This leaves a distribution of x/12 per manifold and energy layer. Then manifold 1 would have x/12 in spin up state and y/12 in spin down state. The same would hold true for manifold 2. The intermanifold layer between 1 and 2 would have a distribution of photons in spin up and spin down as well.

Now, as renormalization theories in quantum field theories postulate, a real photon with spin up enters manifold 1. This skews the spin up portion in manifold 1 .

So, a newly created photon in manifold 1 goes into a correlation spin down state to compensate- or possibly into another spin up if the total is already skewed negative even with the addition of the first one. This will happen regardless of where the photons are created. It is more of a global balancing act for pressure as in our pipe example.

This happens across all the manifolds in a more complex way than I am describing, because the distributions are in flux. There is also the issue that multiple variables like spin, position, momentum and polarization are involved in the state definition.

The automatic correlation is happening not because any of these photons "know" the state of the other, any more than the water molecules in our pipe example know the state of the water molecule at the other end.

It is because the overall states must balance out. The cumulative state of the photons should not change arbitrarily since the Big Bang. The intermanifold energy layers together act as a gatekeeper to keep the contributions balanced out.

This global count and maintenance explanation seems far more plausible than any individual photon to photon exchange of information about their state. The scaffolding of spacetime is like a lattice of observable points, separated by intermanifold energy and quantum folds. As noted, before, the intermanifold energy layer operationally acts as both a gatekeeper and a nonlocal link between particles that may be widely spread out in their local manifolds.

Let us look at this another way. You have teeter-totter 1-6 lined up in parallel, with kids on all of them and a bunch of kids waiting to get on. A kid gets tired on teeter-totter 1 and jumps off and runs back into line, intending to jump back on when he feels like it. One of the kids that is lined up sees that teeter-totter 1 is off kilter and jumps on automatically.

He does not need to know how many kids are on the teeter totter on each side, nor does he need to know the kid who jumped off. All he sees is that it is off balance and he instinctively jumps on because he wants to be part of the game. And so it is with all the teeter-totters.

Through all the fluctuations of them, there is an automatic balancing of weights on both sides through all their activity. There is no mystical knowledge of what each kid is thinking or doing, other than the fact that balance is off.

So it is with our universe. The original Big-Bang universe was pure energy and had an overall balance. As it cleaved off into its constituent pieces, the 6 spacetime manifolds have their fluctuations, but the combined pools of inter-manifold energy plus the contributions from the spacetime manifolds must maintain their balance. This requires activity that looks like correlations across vast separations in spacetime.

G. Wave Particle Duality

It is well known in physics that there is a fundamental duality to nature-sometimes objects on the quantum scale behave as if they were what we would classically think of as particles. At other times, they exhibit distinct wavelike qualities. In a practical sense, this schizophrenic behavior has little effect on calculations. It is more of a philosophical point of how you view the object.

I would like to offer a different vision; the quantum scale object is in fact always a particle in one of the six-spacetime manifolds, while it appears as a wave in the other. It does not ever change its characteristics; it is only our limited view of its action that is at stake.

The following picture demonstrates the connection between the six spacetime manifold model and the view of a particle as a wave:

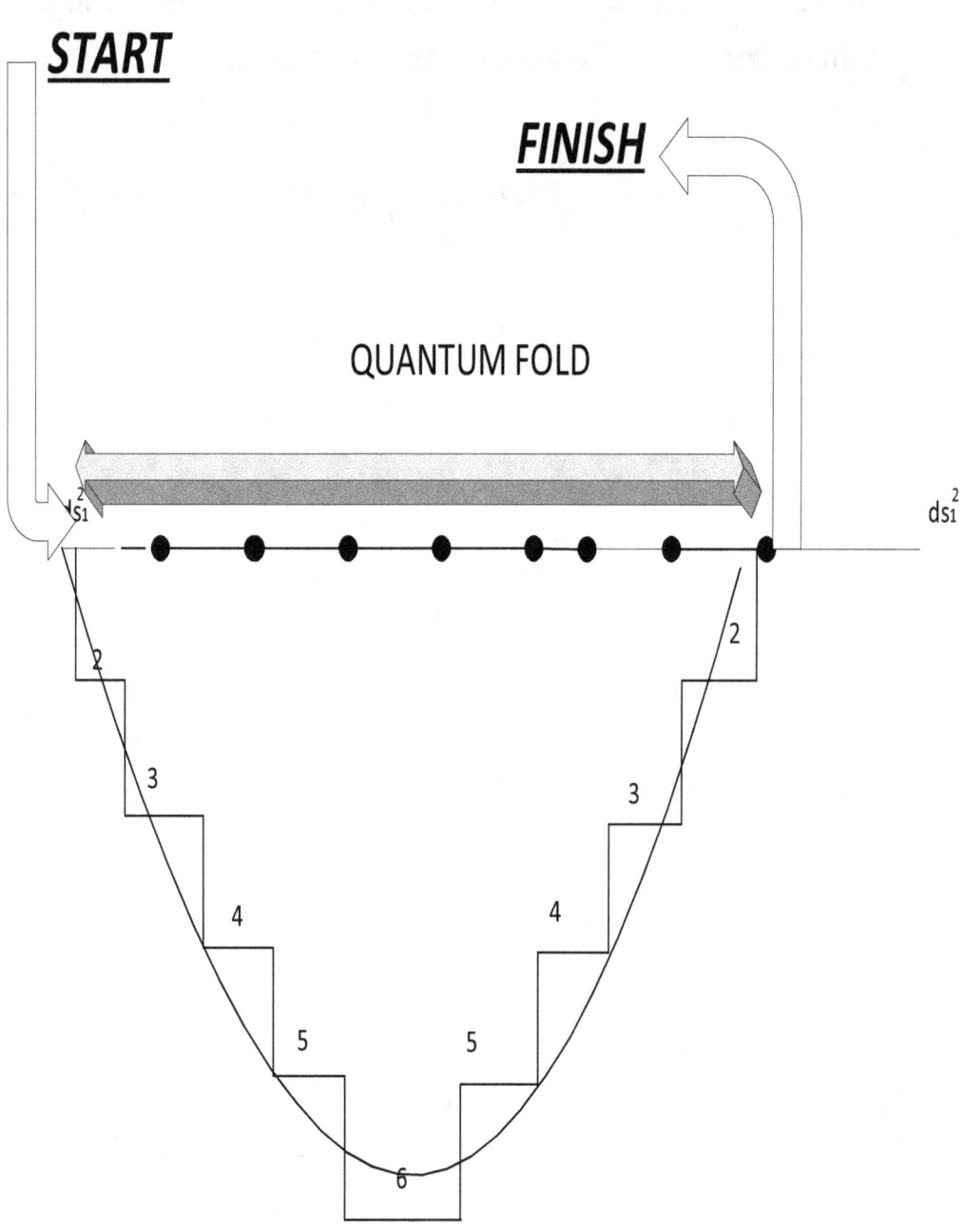

START is in our manifold as is FINISH. Between them lies the chasm of the quantum fold. As the particle falls through the intermanifold energy layer and into manifold 2 , we have no idea where it is. We can draw an arc as in the figure to indicate some probable path. The probability of the particle being in manifold 2 is:

$(imf_{12}ds_2)^2 /((imf_{12}ds_2)^2 +(imf_{13}ds_3)^2 +(imf_{14}ds_4)^2 +(imf_{15}ds_5)^2 + (imf_{16}ds_6)^2)$

The same holds repeatedly for manifold 2-6. I pictured the downward path first and the upward path second, but I could equally well picture it the other way- there is no preference for the direction so long as we have a balance between the "push" and the "pull" . Recall that any given manifold is adjacent to two other ones. For example, manifold 1 has manifold 2 and 6 as its adjacent manifolds. We can think of manifold 2 "below" manifold 1 and manifold 6 "above" it. The corresponding inverted wave is shown below:

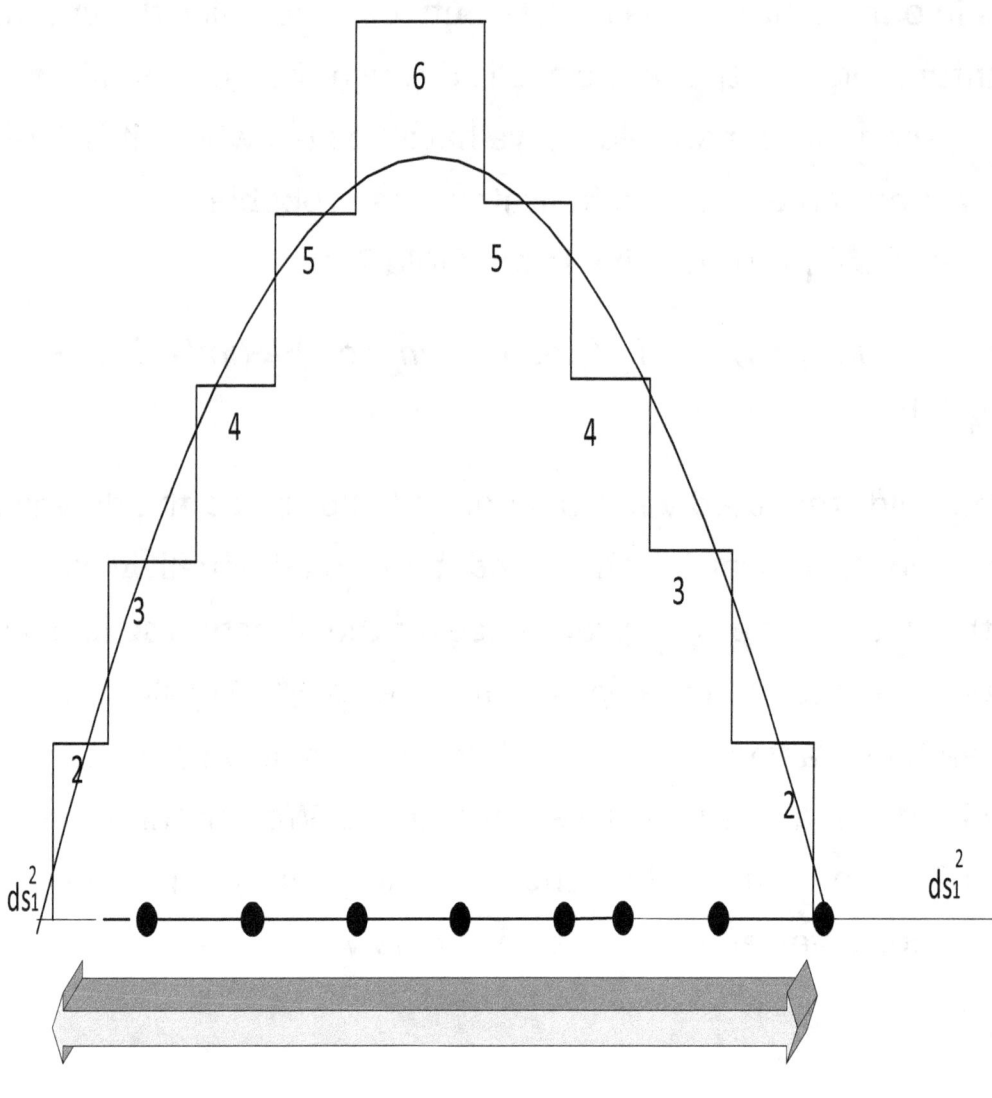

QUANTUM FOLD

Repeated traversals like this across the quantum folds results in:

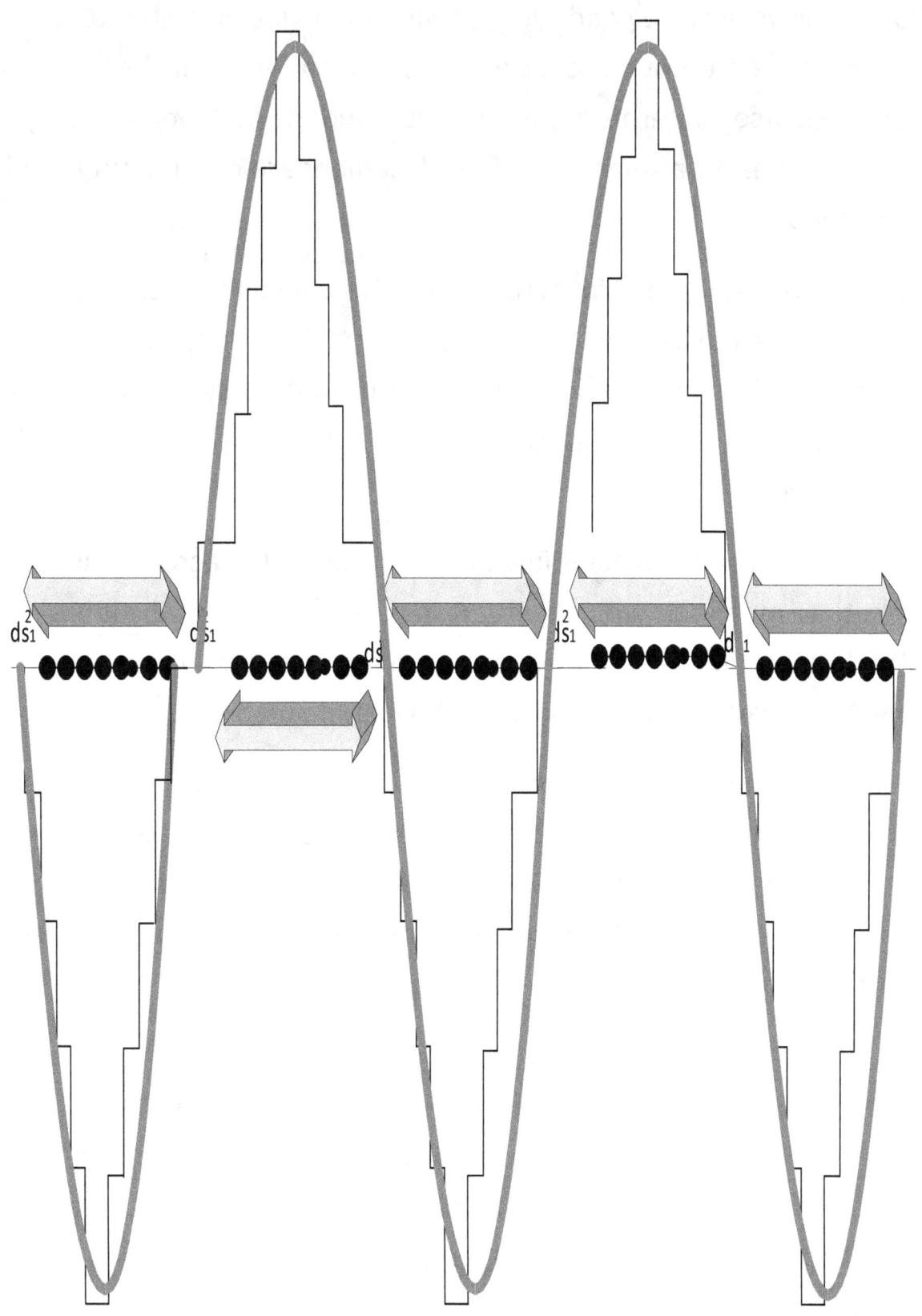

Note how we achieve good synergy with all the ideas in quantum mechanics: representing particles as waves, notions that the location or state casts itself in terms of probabilities, and the inability to measure the properties for a small slice of displacement are natural outgrowths of my model.

The filter of reality that limits us to perceive only one spacetime manifold leads to a distorted view of the quantum scale object changing its form. In fact, it retains its form, and we see only a sliver of its existence. Viewed in this light, there is nothing mysterious about wave particle duality.

The quantum scale object is just happily living its life across multiple spacetime manifolds with observers in each wondering why it is a particle sometimes and a wave at others. You can always view the object as a particle, only in different spacetime manifolds.

As it transitions between them, observers confined to one spacetime manifold must construct a formulation to explain the location of the particle. The quantum mechanical formulation of the entity as a wave that collapses into an observable with probability coefficients works. My model explains why this is the case.

The famous Young's double slit experiment illustrates wave particle duality vividly. This experiment shoots particles such as photons, electrons, atoms, and molecules through a double slit apparatus. As they go through one at a time, the results look as you expect- impacts from single particles appear on a screen on the other side of the double slit apparatus.

As you send many particles through, a wave interference pattern appears on the screen as if the particles were really like waves. This puzzling phenomenon is impossible to explain with any classical models of physics. Quantum physicists cannot really explain it; they just accept it as a model that works.

Let us see how the results of this experiment fit into my model. When you shoot particles one at a time, the resultant path would be transitioning across six spacetime manifolds. The net result of all these transitions would be the particle moving from point A to point B , but concretely appearing in our manifold only some of the time.

As you start sending many particles across, the same transitioning occurs, but there is overlap between particles bumping into each other as they make the transitions. The result is that their emergence at the endpoint produces an interference pattern.

In short, the interference pattern is the result of multiple particles transitioning in and out of multiple space time manifolds at exceedingly small scales, colliding with each other (as in a hydrodynamic fluid or a plasma) and producing an interference pattern.

Just as individual water molecules can become waves in an ocean when combined in a unique manner, individual particles transitioning via space-time manifolds look like waves as they arrive at the common endpoint, colliding and being disrupted along the way.

Here is a good analogy that will clarify why an interference pattern emerges when there are many particles, but not when there is one. Imagine a highway with very deep valleys and hills. A car starts out at one end of the highway (at ground level) and you are observing its progress from the other end- you are also at ground level.

As the car is traveling on its path, you see it disappears and reemerges (as it goes down a valley and up the hill back to ground level). With one car, you can chart its progress and confidently state that it was that car at the end of the journey.

Now imagine you have many cars going through the same journey. It will quickly get confusing, and you must start guessing about which car was arriving. You will find that as you put probable locations of each car, you are ending up drawing a pattern.

In this analogy, the hills and valleys are the hidden spacetime manifolds and the cars represent the particles. This gives you an easy to grasp intuitive picture of the wave-particle duality in the 24-dimensional universe. It lacks collisions along the way, so it is not a precise analogy, but I think it conveys the essential point.

In summary, wave particle duality has a physical foundation. Viewed as a smooth traversal across all manifolds you can observe particle like behavior. Confined to the view of a single manifold, it is best represented as a wave that collapses into certain endpoints that end in our manifold that correspond to points across a quantum fold.

H. Fit with existing interpretations of Quantum Mechanics.

There have been numerous analyses of the foundations of quantum mechanics- too many to cover in depth here. The major ones include:

- ➤ Many worlds interpretation- a mechanistic view of the fundamental equations of quantum mechanics, resulting in a split of the universe into many worlds at every quantum split of the wavefunction. This expansive view includes all objects as part of the overall quantum mechanical wave of the universe, splitting a rather large number of times. Other than the complete lack of intuitive appeal, it appears to be a self-consistent interpretation of the foundations of quantum mechanics.
- ➤ Bohmian mechanics, which suggests a pilot wave and a particle with no reciprocity, where wave provides a guidance equation for the particle. This provides a very concrete foundation for both a particle and a wave view of fundamental entities but suffers from the fact that there is no effect of the particle on the wave. There is also the issue of lack of Lorentz invariance in this view.
- ➤ Decoherence theories, that suggest that spontaneous wave collapse explains the difference between macroscopically stable matter versus quantum particles.
- ➤ The prevailing standard explanation, the Copenhagen interpretation. This is the prevailing "textbook" view every physics student gets in quantum mechanics. It involves probability coefficients that collapse into physical observables after a measurement. The wavefunction describes a superposition state before measurement and collapses into an eigenstate corresponding to a physical quantity after a measurement.

The natural question is : how does my model fit in with these foundational interpretations? My view of all elementary particles when viewed in a single manifold is that of a wave. In this respect, I am I alignment with the view that all objects are best represented by waves. The wavefunction has a more expansive structure, traversing over all six manifolds and intersecting with ours.

The branches that proliferate in the many worlds view are not part of my model. I think the branching , insofar as it may occur, would be over the six manifolds cyclically. Perhaps the large number of branches in the many worlds interpretation could be mapped cyclically to the six manifold model (ie large number of branches reduced to a finite set of cyclical branches , which would also be large in number).

There is also some room to find commonality with the pilot wave view. A particle cycling through the various manifolds can be represented loosely by a wave. The wave could represent the pilot wave that is guiding the particle through the various manifolds, when viewed in one manifold. Viewing the particle over the entire six manifolds would more closely resemble a traditional particle. Here the particle has an impact on the wave because as it falls through the intermanifold energy layer, it contributes to the amplitude of the wave.

In terms of decoherence theories, the idea of coherence in my model needs further examination. My view of coherence in the macroscopic world is that the elementary particles have achieved stability at the endpoints of quantum folds and combined in the appropriate manner to form, for example, a detector.

On the other hand, small scale systems have not yet achieved coherence. They become subjected to the forces involved in the quantum fold and can move forward into multiple branches, leading to the standard superposition states in quantum mechanics. Decoherence is flipped on its head in this type of interpretation.

It is true that macroscopic objects like a detector are composed of smaller elementary particles that are wave like and subject to the rules of quantum mechanics. However, large collections of these smaller wave/particles cohere on the lattice points that rest in manifold 1 . That is what makes a macroscopic object, the coherence of these elementary particles into real points on the lattice that exist in manifold 1.

The substrate of the lattice points is the intermanifold energy and if the constituent particles are there, they are not in manifold 1. If enough of them are not in manifold 1, there will be no coherence and no object present in manifold 1 that can be detected by any means.

In terms of the standard textbook view of quantum mechanics, my model would have the concept of the wavefunction (across quantum folds). The probabilities occur because of the branching the wavefunction may take in terms of its interaction with the energy layer in the quantum fold, as well as exactly which lattice point it achieves after traversal of several quantum folds.

The collapse of the wavefunction with a measured value is more like the intersection of a particle/wave with coherent lattice points. The coherence of any large-scale object has already happened in my model. There is no need for a collapse in the Copenhagen interpretation sense. The entanglement of the microscopic and macroscopic objects is described in terms of interaction between an object that is actively traversing quantum folds and can be described as a wave, and a macroscopic object whose collection of waves (all the quantum pieces that make it up), have already cohered at observable points in our manifold. Coherence as a macroscopic object is a constructive stochastic process across six manifolds for a collection of waves which materialize in our manifold.

I. Vacuum Energy

Another basic feature of modern quantum field theory is that the vacuum is a very lively place. It consists of an infinite sea of virtual particles, which are created and destroyed all the time, producing energy.

These are particle-antiparticle virtual pairs that are apparently everywhere.

There are a couple of fundamental difficulties with the quantum vacuum energy. First, it produces too high a number in terms of any correlation to observable reality or cooperation with General Relativity and the cosmological constant.

Second, we are motivated to ask where do all these virtual particle-antiparticle pairs come from ? There seems to be no reason for their existence other than to make the equations work out.

It is not a comfortable picture of the universe that we introduce a sea of virtual particles that are apparently there but not there. Somehow, they contribute to the vacuum energy, yet they are not there.

I think a more physically meaningful explanation is in order. In my model, manifolds one, three and five are matter dominant while two, four and six are antimatter dominant. This seems logical since the original universe was just pure energy. What we perceive as the vacuum energy is a small funnel that interacts with your antimatter counterpart.

The following diagram shows manifold 1, which is adjacent to manifold 6 and manifold 2. Also shown is the physical origin of the vacuum energy:

Antimatter Dominant Manifold 6

5% visible mass local to manifold

INTERMANIFOLD ENERGY 11.67 PERCENT
ENERGY OF UNIVERSE

Matter Dominant Manifold 1   5% visible
mass local to
manifold

INTERMANIFOLD ENERGY   11.67 PERCENT
ENERGY OF UNIVERSE

5% visible mass
local to
Antimatter Dominant Manifold 2 manifold

The small quantum folds in each manifold are occupied by transitions between manifolds. These are indicated by the lines connecting the manifolds through the intermanifold energy layer.

Following the fundamental assumption of the model, the Big Bang caused a split into six manifolds- three of them matter dominant and three of them antimatter dominant.

Each manifold has a quantum fold through which particles like photons are coming across the manifolds. The pool of intermanifold energy is what is left over from the Big Bang- it is quantifiable, not an infinite sea of virtual particles.

Each adjacent manifold being antimatter dominant in the case of manifold 1, we can expect some antimatter matter reactions on a small scale.

In other words, if you are in manifold one, three or five small tunneling antiparticles from two, four and six are constantly producing lively activity in what we think of as the vacuum. The vacuum consists of an interweaving of all the six-spacetime manifolds. The reason the value turns out too high is that we are not accounting for the picture in the other manifolds and assuming the energy ends up all in our native spacetime manifold. A significant portion of this energy will be in an intermanifold transient state and not contribute to the total in our spacetime manifold.

Subtracting that value off will leave us with a net that will make more sense in terms of corresponding to observable reality and how it fits in with the cosmological constant and General Relativity.

To summarize, these virtual particle-antiparticle pairs are not virtual at all! They are very real particles contributing the quantum vacuum energy, with some energy trapped in intermanifold layers and some energy being really produced in the native spacetime manifold. Adjusting the formulae to consider these factors, the quantum vacuum energy will end up being a more sensible number. The quantum field theoretic number is an earnest one, adding up all the intermanifold energies. The real number is tamed by the size of the quantum folds, which allows a very tiny portion of this energy to leak into our native manifold. That is the mathematical adjustment that is needed to bring quantum field theory and the cosmological constant into synchronization.

J. Renormalization

Let us consider some of the scenarios that lead to the need for renormalization. The first is when a photon creates a virtual electron-positron pair that annihilate each other. The question is what is the source of this "virtual" pair? In fact, why are they virtual at all? Are they just real particles that we have failed to incorporate into the theory?

The answer in my framework lies in the cross-manifold interaction. What we perceive as a virtual electron-positron pair annihilation is in fact a crossover of an electron in manifold 1 (where matter is dominant) to manifold 2 (where antimatter is dominant) and really annihilating themselves to produce some energy.

The first situation is depicted in the following diagram. The particles are real- there are no virtual particles. If they transit to another manifold, they are classified as virtual. There is an ocean of intermanifold energy, constantly churning particles in and out between manifolds.

RENORMALIZATION SCENARIO 1 ENLARGED PICTURE

Photon

Quantum Fold

MANIFOLD 1

MANIFOLD 2

Quantum Fold

Quantum Fold

Electron

Energy

Positron

This takes the "virtual" out of virtual electron-positron pair annihilation but leaves the entire existing math intact. We are just reinterpreting the math to include real particles in other spacetime manifolds. The interaction could happen in the intermanifold region as well. Picturing it in manifold 2 gives it a more concrete foundation for visualization physically. Dark energy, in my model, is energy trapped in the intermanifold region. The energy interactions involving virtual particles in renormalization scenarios are coming from this region.

The second scenario is when an electron emits and absorbs a virtual photon, leading to a self-energy condition. Once again, why is the photon virtual? Is it in fact real and why do these interactions occur? What is the underlying physical process/structure that makes this happen?

The second scenario is depicted below in my model:

RENORMALIZATION SCENARIO 2 ENLARGED
PICTURE

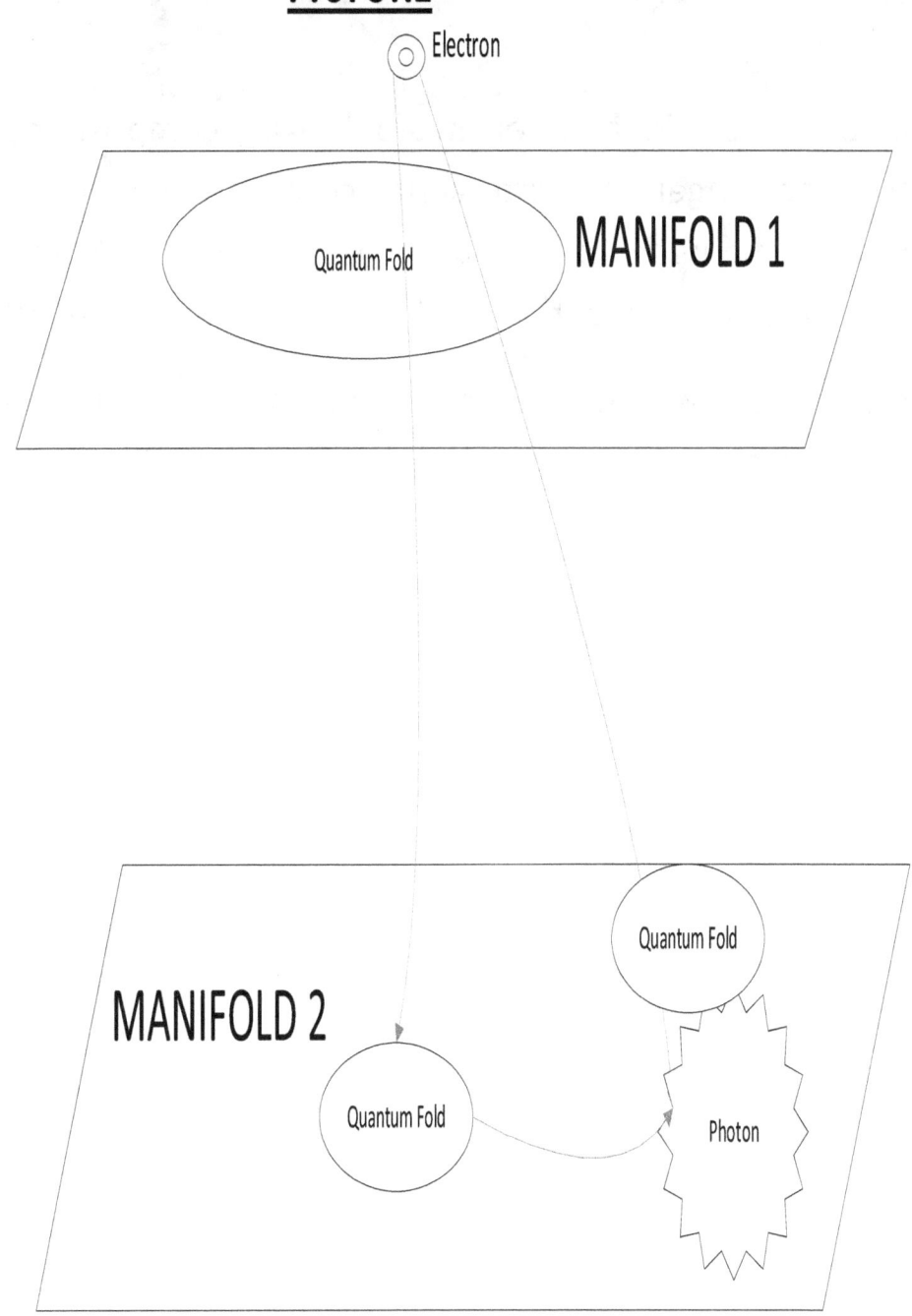

The interpretation is straightforward. The photon is a real one; it gets emitted and makes a transition to an adjacent manifold. It makes a trip back to ours and affects a local electron. There is a continual dance of a small number of photons transitioning between the various spacetime manifolds, which to us, looks like they are appearing and disappearing from nowhere.

The ocean of photons in the intermanifold region compose the layer we call "dark energy". Again, this interaction could happen straight out of the intermanifold energy region, but the visualization of the interaction is so much clearer when we think of the interactions in manifold 2.

The third scenario is where an electron emits a photon, then emits a second one, and subsequently absorbs the first one. This situation is depicted below:

RENORMALIZATION SCENARIO 3 ENLARGED
PICTURE

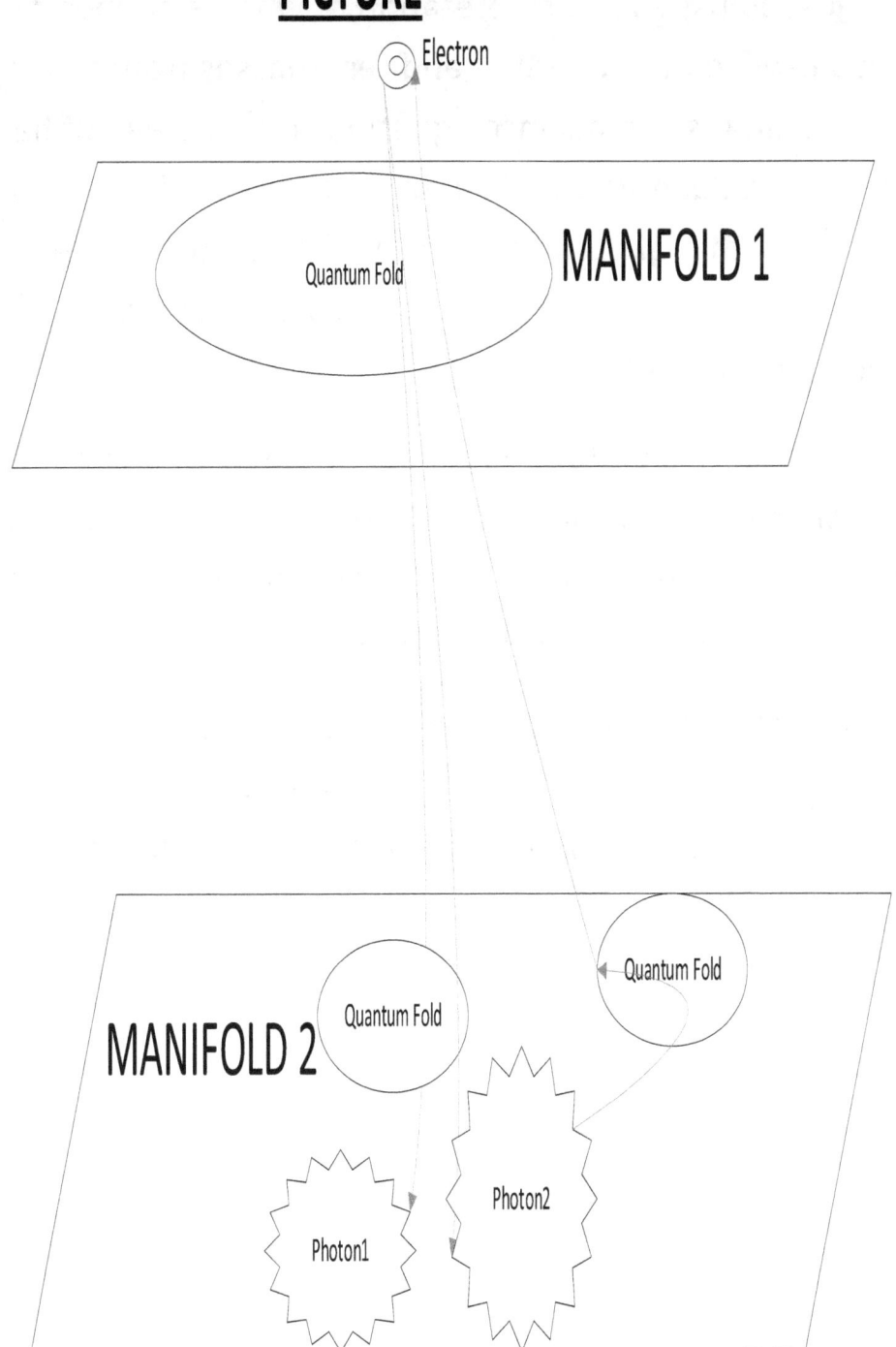

This reservoir of appearing and disappearing photons are crossing over from another spacetime manifold in my model.

When you just adjust your interpretation of what is happening here slightly, you can factor in energies and rest masses from other spacetime manifolds and balance equations out, instead of having to introduce infinite quantities and subtract them. The bottom line is to replace virtual with real particles and carefully factor in energies from the intermanifold energy layer and rest masses that are resident in other spacetime manifolds.

This will replace the need for infinite quantities with a finite pool of particles residing in adjacent spacetime manifolds. We need to be careful about sizing the intermanifold layer and not double counting particles since there are interactions between manifolds.

The easiest illustration to see how renormalization theories can be positively influenced is to review the self-energy of an electron as a qualitative calculation in my model. The electron looks as follows:

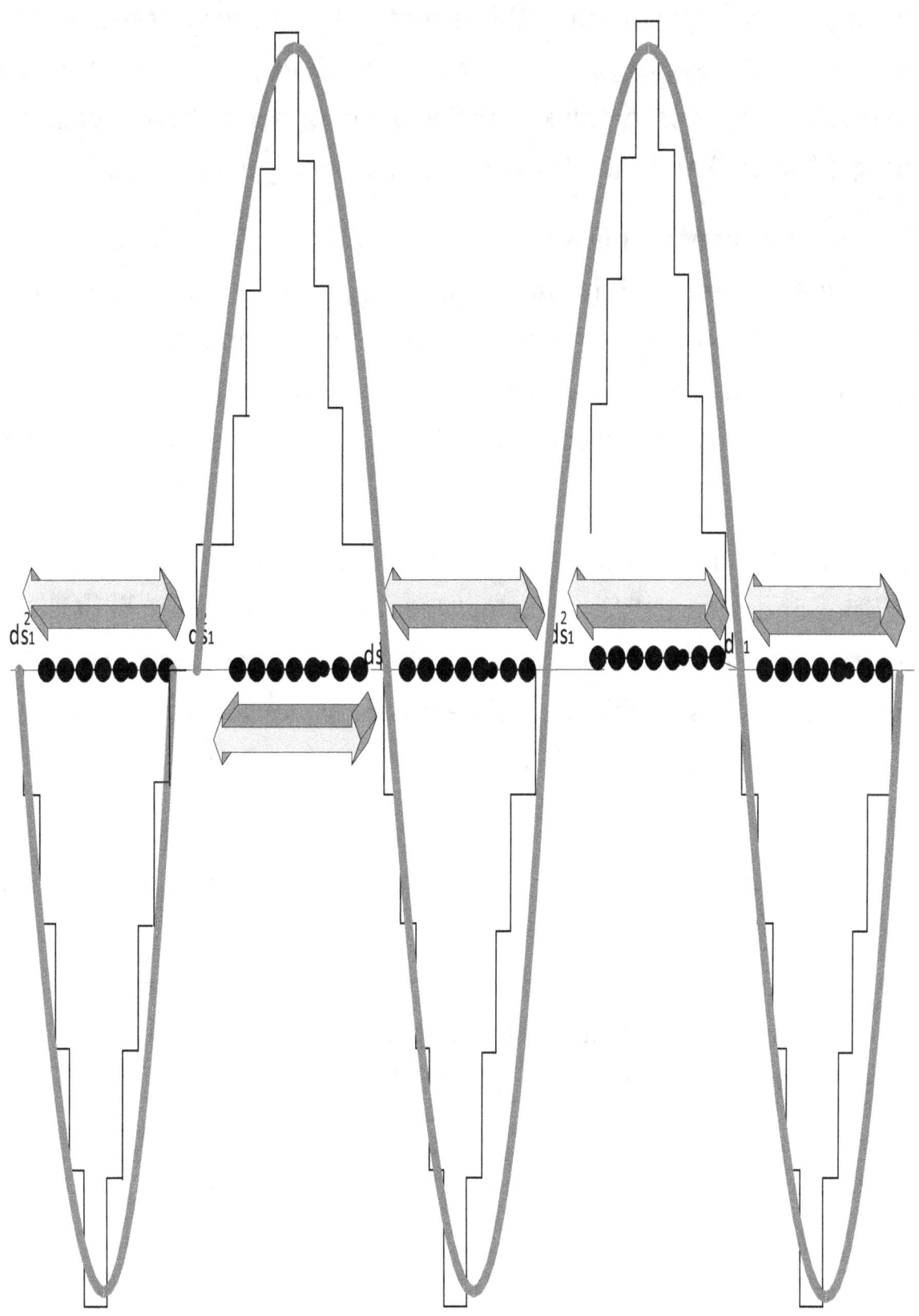

The physical interpretation of this is that as the electron traverses the intermanifold energies surrounding the manifolds, it picks up and loses energy. You can loosely think of the travel in one direction as gaining energy("down") and the other direction as losing energy("up").

The overall energy of this wave, like any wave is dependent on the amplitude squared and the wavelength. In particular, the amplitude is related to the journeys through the intermanifold energy layers. The further it travels and interacts with the intermanifold energy layers, the larger the amplitude and consequently the higher the mass/energy of the electron.

Recall that the intermanifold energy layers will consist of an ocean of particles and antiparticles, which will accompany the journey of the electron. The self-energy of the electron is a finite one, because as in any wave, there is nullification of peaks and valleys and the overall energy ends up dependent in a large way upon the amplitude and the wavelength.

Considerations like this will result in a more realistic quantum vacuum energy calculation and a more realistic explanation of why we are renormalizing equations. What appears to be an infinite source of virtual particles are in fact a finite source of particles getting distributed between our six spacetime manifolds. The fundamental wavelike nature of all the elementary particles will nullify the infinities that tend to creep in when you start looking at all virtual interactions.

K. Mysterious "hidden" dimensions.

Theories attempting to unify the various field theories introduce "hidden" dimensions- spatial dimensions that are curled up and invisible in such a tiny piece of space that we cannot possibly see them. The idea of introducing extra dimensions to try and make the theories consistent is a good one.

The most popular version of adding dimensions is string theory and superstring theory. The following chart demonstrates the zeal for adding various numbers of dimensions.

Theory	Dimensions
Bosonic (closed)	26
Bosonic (open)	26
I	10
IIA	10
IIB	10
HO	10
HE	10
M-Theory	11

This zeal is prompted by a very earnest desire to create a unified field theory that explains everything. *But it violates the spirit of relativity to asymmetrically introduce these as spatial dimensions only*. Recall again the famous quote by Minkowski:

Henceforth space by itself, and time by itself, are doomed to fading away into mere shadows, and only a kind of union of the two will preserve an independent reality."

Consequently, if you are going to look at fracturing, extending, or hiding dimensions, you are compelled to look at hidden spacetime dimensions. It is counterintuitive to go and hide spatial dimensions without hiding time dimensions along with them if you fully embrace relativity.

Based on the amount of 'missing' mass in the universe, I constructed five hidden dimensions in a sense. It is just that instead of adding them just as spatial components as string theorists do, I added them as five distinct spacetime manifolds superimposed on each other with distributed mass. This is depicted in the diagram below:

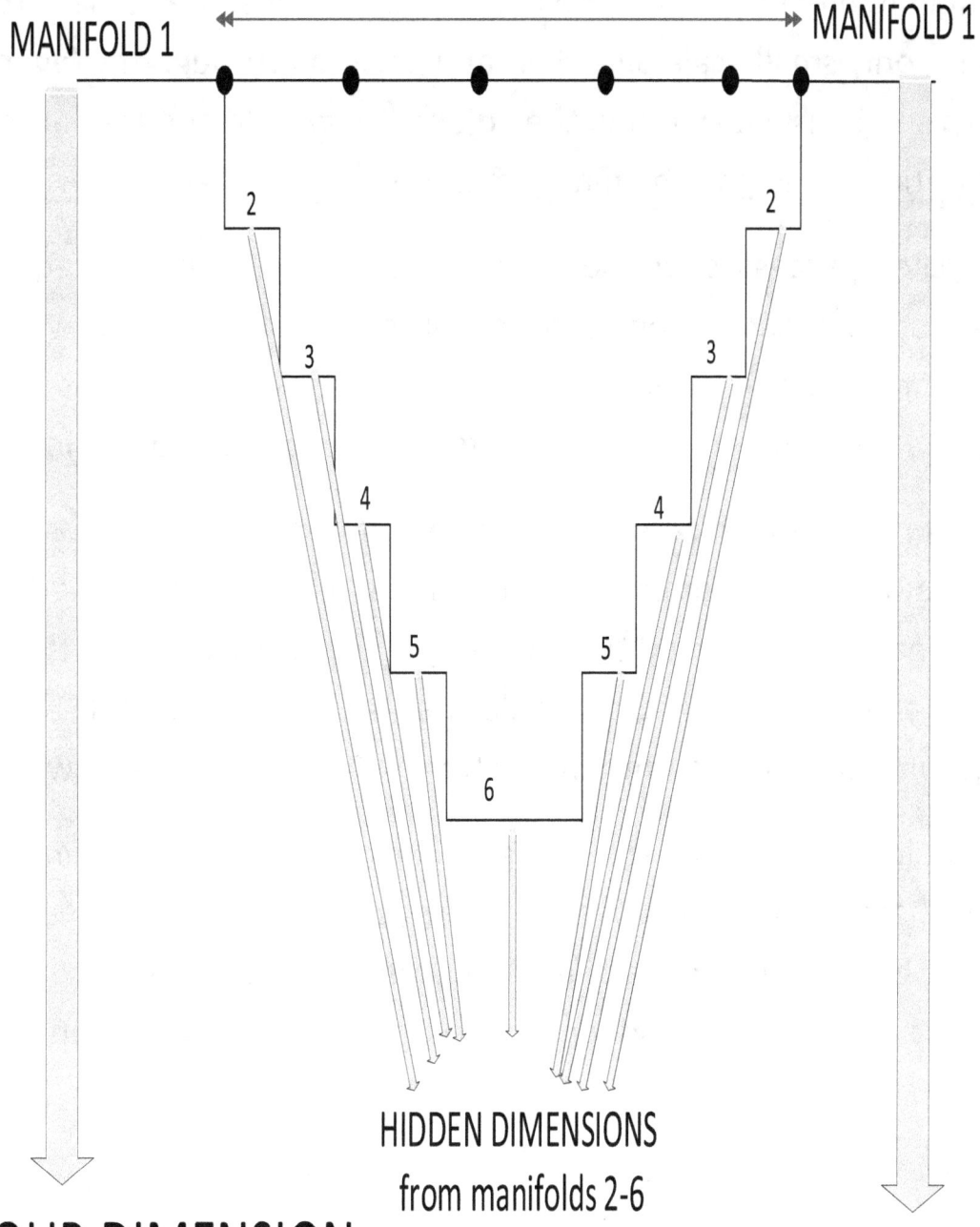

QUANTUM FOLD

MANIFOLD 1

MANIFOLD 1

OUR DIMENSION
Manifold 1

OUR DIMENSION
Manifold 1

HIDDEN DIMENSIONS
from manifolds 2-6

Note that each spacetime manifold is separated from the next by an intermanifold energy layer. The hidden dimensions look "small" to us because only small-scale particles can transition through the layer, make a small displacement in the adjacent manifold and repeat this cycle until returning to the native manifold.

This cyclical process is symmetric to every manifold. The consequent nearly continuous 4-dimensional spacetime manifold is what we understand the laws of General Relativity to apply to on a mass scale. On a small scale it very naturally suggests a quantum mechanical world.

Let us consider why it makes sense to add spacetime manifolds as hidden dimensions- I maintain that this is at least as valid as adding purely spatial ones. I need to take a slight detour to illustrate this. Imagine the speed of light is 60 miles per hour instead of 186,282 miles per second- a rather astonishing factor of 12 million times slower.

You would notice some very strange things. First, you would find it awfully hard to move because the energy required to move even at 1 mile per hour would be massive. This is because the higher order terms in kinetic energy would suddenly become very prominent even at low speeds.

But since we are just imagining scenarios, let us imagine that we can move around like the way we do now, without violating the absolute 60 miles per hour speed limit. Then if you walk around the neighborhood and come back home, you will notice that your clock and your wife's clock no longer agree.

Driving a nuclear reactor powered car (remember kinetic energy is a lot harder to come by!) across the interstate at say 30 miles per hour, you would notice that the yellow field looks blue out your front window. But if you look in the rearview mirror, the same field looks red.

If you manage to reach a speed of 54 mph by pumping enormous energies into your car engine, you will notice that everything in front of you seems to have shrunk by about a factor of 2. For you, your heartbeat (natural clock) and any clocks you are carrying tell you time is proceeding normally.

Someone watching from the road would say that your clock slowed down by a factor of 2! While 1 minute was passing for our observer at rest, it would look like only 30 seconds passed for you. One interesting thing is that you could live longer (at least from the perspective of the observer) by simply traveling fast relative to the 60 miles per hour limit.

Rich people might pay for the privilege of time dilation like some do now for cryogenic freezing. There would be other effects such as any signal that now traverses the earth in milliseconds would start to take hours (television signals, for example).

Any theoretical physicist or ordinary person living in a hypothetical universe like this would instinctively come to the correct conclusion that it is much more accurate to view spacetime as one combined dimension where each affects the other. It is not the conclusion we come to instinctively only because the speed of light is so high. Luckily, the genius of Einstein penetrated the fog of appearances and concluded the true nature of spacetime.

Having finished my detour- what does all this have to do with hidden dimensions? Well, if you asked a theoretical physicist in our imaginary universe how he/she would add hidden dimensions to complete theories, he/she would answer "well of course we should add hidden spacetime dimensions". They would consider this as much a "common sense" approach as we do adding purely spatial dimensions to make theories work.

I think this shows that the bias towards adding hidden dimensions as spatial dimensions is purely based on our terrestrial experiences. I instinctively think of space and time as distinct dimensions, as everyone else does. When I am confronted with physical phenomena that seem to be hidden from me, my instinct is to add a physical dimension to explain the discrepancy. It is easy to visualize, for example, a two-dimensional world, where adding the third dimension resolves apparent paradoxes. This is tied into our instinct of viewing space and time separately. Consequently, when confronted with seemingly paradoxical physical phenomena, our instinct leads us to add spatial dimensions to resolve them.

If I lived in a universe where the speed of light is low and I instinctively realized that space and time are a combined dimension, and I am confronted with physical phenomena that seem hidden from me, my instinct would be to add another combined spacetime dimension. This is closer to the real universe, and hence should be the approach we adopt.

I believe our bias has robbed us of an opportunity to explore the true nature of hidden dimensions. Our colleagues who lived in a parallel universe where the speed of light was sixty miles per hour would be instinctively biased the exact opposite way, adding spacetime manifolds!

This approach provides a high-level unification of relativistic and field theoretic approaches. I can see no reason for discriminating against relativity when adding dimensions - let us be democratic in our expansion of dimensions and abide by the rule that spacetime is really a block dimension!

What is easier to grasp and makes more consistent intuitive and theoretical sense? Extra spatial dimensions curled up invisibly introduced solely for making equations work out or extra spacetime manifolds that are interwoven with ours? This is a simple question for me- it is obviously the latter. The rationale for me that tips this in favor of spacetime manifolds is that extra spacetime manifolds are physical constructs. Moreover, they are physical constructs with physical laws that are essentially identical to ours. This removes the mysterious aspect of hidden dimensions and grounds it back in physical reality.

L. Relationship to the Standard model, String theories and Loop Quantum Gravity

I have not yet discussed any existing unified field theories, except briefly in Section G. Nor have I discussed the Standard Model in particle physics. It is time to remedy these deficiencies.

The Standard Model divides the family of particles into leptons, quarks and bosons. The family of leptons consists of electrons, muons, tau particles, electron neutrinos, muon neutrinos and tau neutrinos.

The family of quarks includes particles of fractional charges of either 2/3 or -1/3 and include down, up, strange, charm, bottom and top.

Finally, the family of bosons is split into the gauge bosons and the recently discovered Higgs boson; the former class consisting of the photon, gluon, W-bosons, and Z-bosons. The latter is a special type of boson that is believed to give rise to the very notion of mass.

A significant point is that the Standard Model views elementary particles as point particles; this has great practical utility and has led to the discovery of several crucial elementary particles using particle accelerators.

In sharp contrast my model proposes that an elementary particle is subjected to the vicissitudes of the intermanifold energy layer and the cyclical travel through the manifolds.

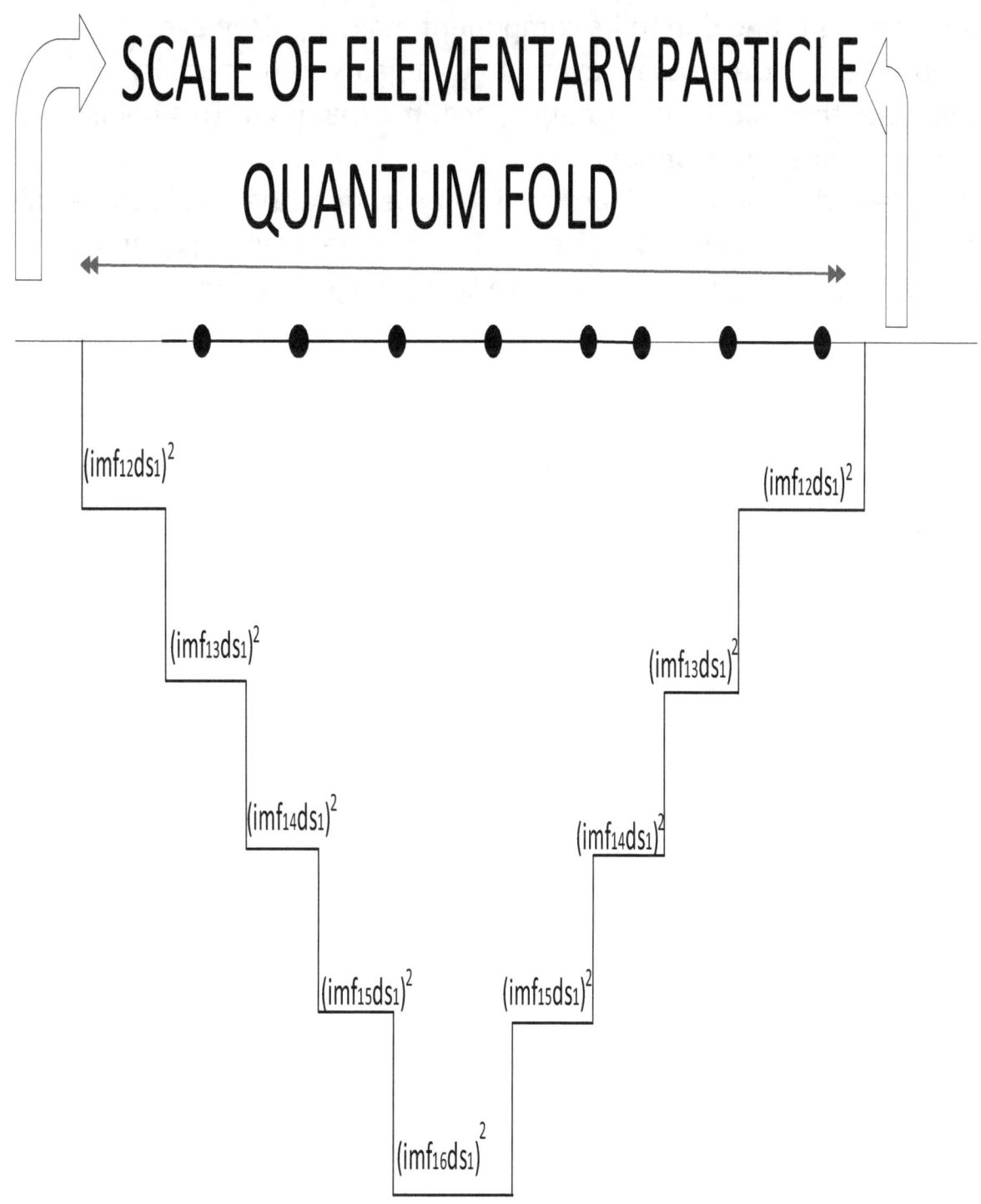

Any elementary particle will traverse at a minimum one quantum fold and is subjected to the uncertainty in its position and energy. This is in line with quantum mechanical expectations. A composite particle like a proton (made up of quarks) will be subjected at a microscopic level to the same buffeting due to its component make up. However, as a composite particle crosses several folds it gains some stability and comprises the visible mass of our spacetime manifold. This holds true for each spacetime manifold.

As explained in the Wave/Particle duality section, these fold traversals result in the particles looking very much like the waves in quantum mechanics. For easy reference, here is the diagram again:

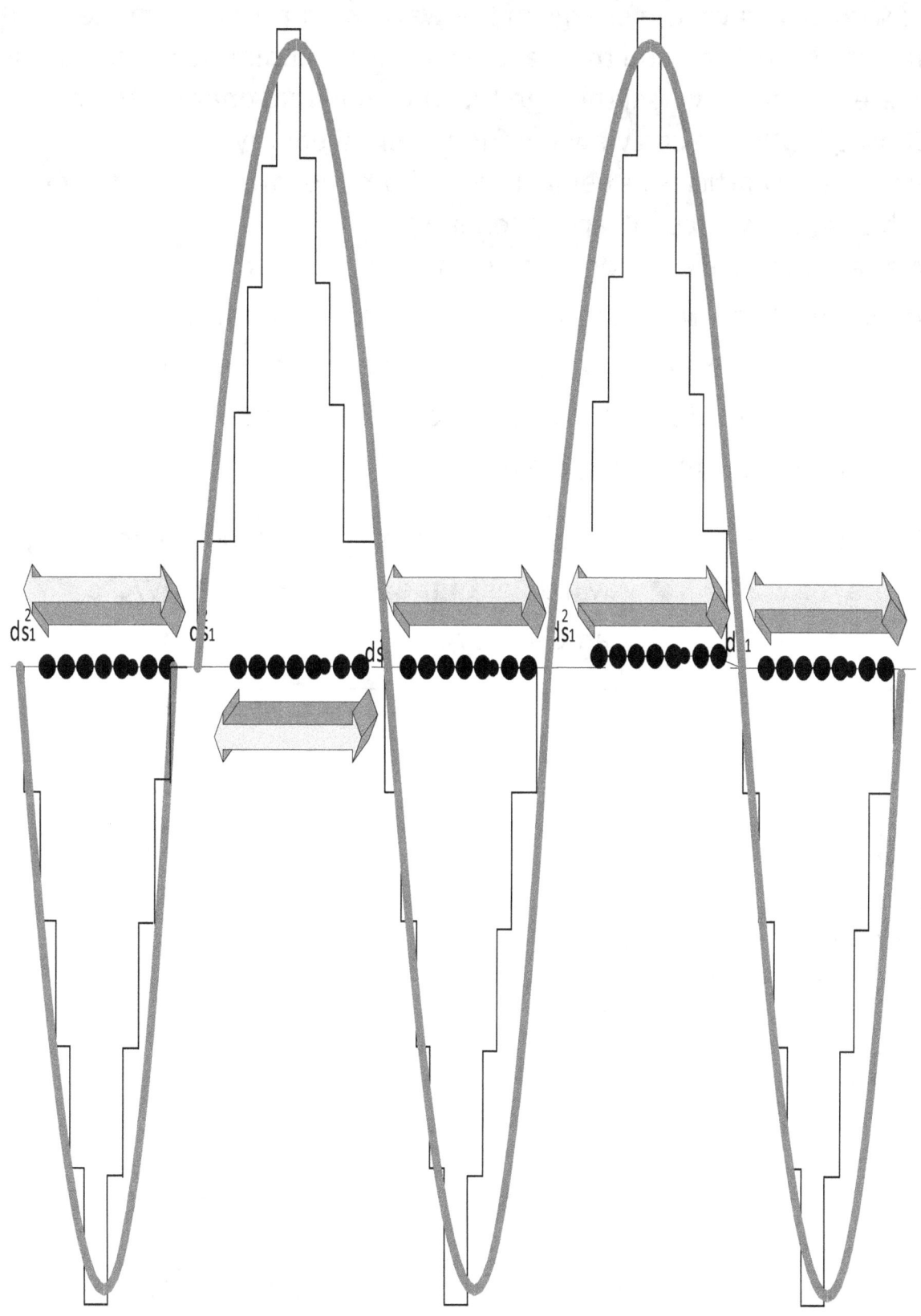

As with any wave, the energy of the wave is proportional to the amplitude squared and the wavelength. The amplitude in this case is related to the traversal through the intermanifold energy region. Consequently, we can view the fundamental energy/mass of elementary particles as being derived from the intermanifold energy (which is the source of vacuum energy).

We can view the electron mass/energy as a fundamental fixed amplitude A and wavelength λ. Interestingly, we can then propose that the remaining elementary particles are multiples of this basic amplitude, guided by estimated values in the standard model, as illustrated in the following table:

Particle	Proposed Integer Multiples of the amplitude A	Proposed Resulting Mass MeV/c*c
Electron	1	.511 (observed)
Up quark	2	2.044
Down quark	3	4.6
Strange quark	13	86.36
Muon	14	100.16
Charm quark	51	1329.11
Tauon	59	1778.79
Bottom quark	91	4231.59
Top quark	581	172493.67

Qualitatively, the emergent picture is that all elementary particles derive their energies from the intermanifold layer. The farther they penetrate the layer , the greater the amplitude of the associated wave and the more mass/energy they gain. This seems to be a reasonable physical proposition. Obviously, the integer multiples were guided by existing estimates of the mass of these particles, but the fundamental point is varying the amplitude gives rise to elementary particles with different masses in this model.

Note that this is not meant to be a comprehensive exploration of the tie between the model and elementary particles. There are far more complexities of the particles involved that are not explained by this simple analysis— however, the fundamental notion of mass seems to have a very straight forward explanation.

The other key properties of elementary particles like spin, charge, color etc, still need to be explored and explained in my model. But giving a clear explanation of mass/energy seems to be a healthy start to me, as well as getting away from the notion of a point particle.

The notion of a point particle creates many difficulties, and my model gravitates very naturally towards the look of an elementary particle as a wave. The phase of the wave and the wavelength perhaps correlate to some of the other key properties. There is scope for using a combination of wavelength, phase and amplitude in combination to find some deeper relationships between the elementary particles.

The underlying geometry is a little more complex than the diagram of the wave makes it appear. The displacement of the particle in each manifold is through a space that conforms to the restricted Lorentz group – SO(3,1) combined with a scalar drop through the intermanifold region , for those familiar with group theory. This space can also be represented as SU(2)xSU(2).

The standard model is represented by the group SU(3)xSU(2)xU(1). The number of generators (fundamental matrices that generate the groups) for SU(3)xSU(2)xU(1) is 8+3+1=12 . Each of these generators corresponds to physically identifiable quantities (8 for the strong force, 3 for the weak force, and 1 for the electromagnetic field).

In my model, the underlying group is SO(3,1) for each manifold. This algebra has 6 generators, combined with drops through the intermanifold energy layers. So, for the overall space we have SO(3,1)xSO(3,1)xSO(3,1)xSO(3,1)xSO(3,1)xSO(3,1)- a total of 36 generators. A better representation might be SU(2)xSU(2) six times over, in view of the key role SU(2) plays in the standard model. Some devilishly clever mapping could associate twelve of these generators with the 12 from the standard model, providing a unified field for weak, strong and electromagnetic field. That leaves some "extra" generators to get gravity into the picture and perhaps account for some other hitherto unknown symmetries and forces.

This is just a conjecture, but it is a possible approach to unifying fields in this framework. The last great attempt at a grand unification theory involved the group SU(5), which has 24 generators. This attempt at unification is very elegant and consistent mathematically- unfortunately, no experimental verification was achieved for it. This gives me hope that another devilishly clever attempt might be to try and map these 24 generators as a subset of the overall 36 in my model and take another run at an SU(5) type grand unification theory.

Another useful comparison is to the promising grand unification theory that relies on SO(10), which has more generators than the basic structure here. It has 45 generators, which is 9 more than what we have here. All these comparisons seem to indicate that any attempt at unification that is attempted based on my model will have a different flavor than the attempts thus far.

Now let us turn to string theories. To reacquaint myself with string theories, I went through the excellent book "An Elegant Universe" by

Brian Greene. I recommend this book highly for its clarity and breadth of coverage.

The gist of string theory is that matter is made up of strings that are on the order of the Planck length in an 11-dimensional space. It is a fascinating theory that is extremely complex mathematically. Note that this theory involves hidden dimensions that are curled up so finely that they are not observable. In addition to three spatial dimensions and one time dimension, there are seven theoretical dimensions.

To make these theories compatible with the observed universe, the notion of compactification of these extra dimensions is necessary. The idea is that all these extra dimensions are so tiny and curled up so tightly that they are unobservable.

All theories that try to extend spatial dimensions to try and unify fields run into similar problems. Their very nature demands that they be physically unobservable on larger scales, and that poses instability as you shrink them. These types of theories have been around in physics for nearly a hundred years, starting with the great mathematical physicist Theodor Kaluza.

They all share the same drawbacks, despite their ingenuity and undeniable mathematical elegance. Whenever you add a dimension, it must be dynamical to fit in with relativity. But that contradicts the fact that it is invisible, so you try to find methods to render it compact. That introduces instability and disqualifies the theory. These theories are balanced on the tip of a needle surrounded by compactification to hide them in one direction and visibility in the other direction.

The more dimensions you add, the worse the problems get. Addition of dimensions just compounds all these difficulties to more dimensions and requires further sophisticated compactification methods.

My model shares the notion of hidden dimensions and I would argue it is fundamentally six dimensional in nature because I view space and time combined as one dimension.

My model amounts to one spacetime dimension we are familiar with and five hidden spacetime dimensions. I hide these dimensions in plain sight, in a manner of speaking. The hidden dimensions are continuous. There is no need to shrink them as they are inaccessible by design. This removes a lot of the traditional problems with hidden dimensions.

The connection between our spacetime dimension with the other ones is through the quantum fold, which I view as the fundamental construct of the universe. The fold size is constrained by a combination of their total displacement multiplied by intermanifold factors and is approximately equal to Planck's constant in magnitude (when squared). Their fundamental dynamism and continuity augers well for any attempt to unify gravity with other forces, a key inhibitor in many attempts at unification.

I view the quantum fold in my model as analogous to the strings proposed in the string theory model- the key difference being the fold protrudes into distinct spacetime manifolds, which are physical constructs. The fact that I am proposing extra physical constructs means there is no need to compactify theoretical dimensions.

The advantage of my model is that our spacetime dimension can be viewed as an embedding directly in a larger construct. Each dimension is symmetric, all are individual spacetime dimensions embedded in a large six-fold spacetime super manifold (or 24-dimensional if you want to count every dimension of space and time).

This model seems so beautiful and symmetric, and I am sure there is a mathematical system that will reflect this beauty. The super manifold is a direct sum of submanifolds with minute overlapping areas that connect the submanifolds.

Due to the presence of multiple extra dimensions, albeit in different form, I wonder if we can achieve the kind of brilliant successes string theory has had and leverage them in my model. This is just idle conjecture, but as string theory succeeded in unifying all elementary particles and forces using vibration of strings, there is the possibility that a similar feat can be achieved in my model with the vibration of a wave perhaps.

Similarly, as string theory gives us all the gauge fields responsible for electromagnetism and nuclear forces using vibrations of open strings, perhaps there is some scope for a similar feat in my model with vibrations of "open waves" or some such construct.

One of best results of string theory is to give rise to gravitons as vibrations of closed strings. Maybe there are "closed waves" or some exotic construct like that in my model that can explain gravitons.

Finally, supersymmetric string theory unifies all forces and particles , unifying bosons and fermions, by describing them as oscillations of strings. Perhaps we can borrow this idea and introduce oscillations of the wave.

In summary, I am hopeful (and it is just a hope at this point), that my model can duplicate the brilliant successes of string theory and avoid its pitfalls due to the fundamental dynamism and continuity of the quantum fold.

Another approach to unifying quantum mechanics and gravity involves loop quantum gravity. It is an exquisitely detailed mathematical theory, involving an atomic structure to space that is dynamic. It is hard to do this theory justice without delving into much detail. The excellent book by Lee Smolin "The Three Roads to Quantum Gravity" has a superb, understandable summary.

One of the basic precepts of loop quantum gravity is that space is not infinitely divisible. This lines up with the fact that in my model, the quantum fold has finite size. You can never shrink the displacement down to zero due to the finite size of the quantum fold. This size, however, is dramatically larger than in loop quantum gravity, which sets limits based on the Planck length, which is on the order of 10^{-35} m, and the Planck area which is on the order of 10^{-70} m^2 . Despite this difference in sizes, perhaps there is some scope for borrowing some ideas based on the fundamental idea of the atomic structure of space.

One of the other features of loop quantum gravity is that it is background independent. It is a theory that holds that there is no spacetime, that General Relativity is more of a theory about relationships between events.

This differs from my view of having parallel spacetime manifolds. On the other hand, having a lower limit for size inside each manifold seems to be aligned with the notions of quantum loop gravity theories. My hope is that some of the ideas from quantum loop gravity can be ported over.

I can do neither string theory nor quantum loop gravity theories justice in this short section. I am only notionally indicating how their ideas may fit in with my model. Curiously, it appears that the basic assumption of a finite distance (quantum loop gravity) gets combined with structure to the basic particles as a wave (analogous to particles as strings in string theory) in my model. Perhaps that is a sign of movement in the correct direction.

M. Mass Distribution in our Galaxy across Six Manifolds

We can use my proposed model to explain the long-standing problem of uniform speed of rotation of stars around the central core of the galaxy. Note the following:

- Stars near the center of the galaxy are moving in nearly circular orbits around the galactic center. The center is now thought to be a supermassive black hole.
- Stars in the halo do not have nearly circular orbits; they have elongated orbits.
- The high orbital velocity of the stars is presumed to be due to the presence of dark matter, possibly made up of neutrinos, WIMP

(weakly interacting massive particles), or MACHOS (Massive Compact Halo Objects).

➤ To account for the high orbital speeds of the outermost stars, you need to have a total mass of about 1 trillion suns inside a sphere encompassing the galaxy. Visible mass only accounts for about 10% of this total.

In my proposed model, we can explain this as follows:

➤ Like a time-lapsed photograph, concentric spheres of masses exist in each of the six space-time manifolds as you move out radially. Each spacetime manifold has its own galactic clusters and local groups.

➤ The orbital velocity staying constant is explained by the additional mass provided by the displaced concentric spheres, along with muted inverse power effects for the other spacetime manifolds. This also explains the lack of circularity in the halo star orbits.

➤ Multiple galactic centers exist, but only one is visible to us. The ones in the other spacetime manifolds are causing perturbations to the halo stars and skewing their orbits.

➤ The mass inside any orbiting star = Mass inside a sphere surrounding the star in spacetime manifold 1 + Mass inside a sphere surrounding the star in spacetime manifold 2 + …. + Mass inside a sphere surrounding the star in spacetime manifold 6.

➤ The first order effect, which dominates the inner stars, is from our own galactic central mass- this extends to about 30,000 light years from the galactic center.

➢ The second order effects skewing and keeping the halo stars from flying away is from all the other mass. We can hypothesize based on observation that the next spacetime manifold has a galactic center about 15,000 light years from our center. In increments of 15,000 light years, each space-time manifold has another galactic center. This would explain the extent of the halo effect, which seems to affect objects 1/3 of the way to the Andromeda galaxy.

➢ The distribution of the mass in the other spacetime manifolds extends far from the center of our own galaxy, explaining the high orbital speeds of galactic clusters.

Here is a cross sectional visualization of how the Galaxy looks in my proposed model.

Like layers of an onion, six different spacetime manifolds are superimposed on each other. One of these is the galactic disk we visibly observe. The rest are what keeps outer stars in stable orbits at higher-than-expected speeds. The easiest way to visualize this is to flatten out the disks to circles for ease of understanding and make the origin of the coordinate axes the center of our galaxy.

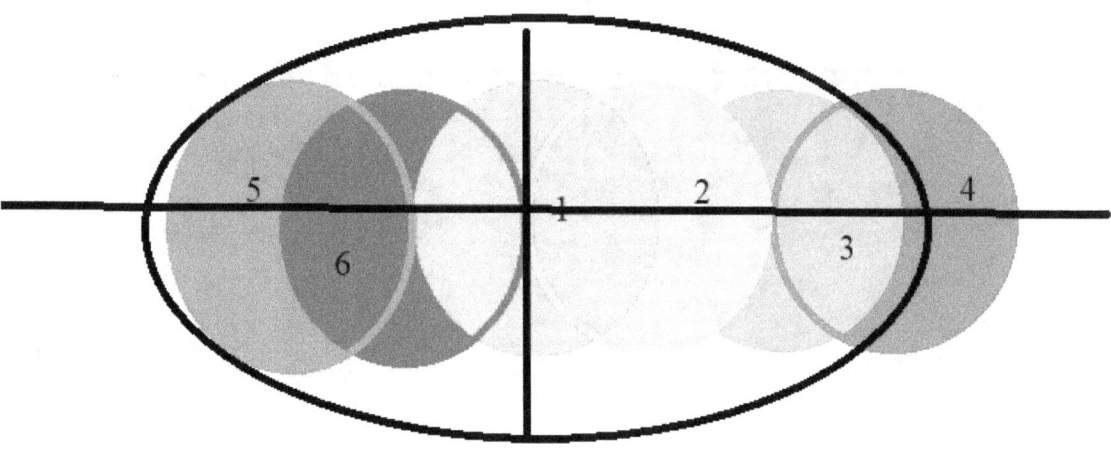

The innermost circle covering a diameter of 30,000 light years represents our spacetime manifold (manifold 1). Each circle represents a galactic core in a different spacetime manifold. The ones to the right of manifold 1 represent spacetime manifolds 2, 3 and 4, respectively.

The ones to the left represent spacetime manifolds 5 and 6, respectively. Every 15,000 light years along the axis represents the central point of a galaxy in a different space-time manifold. *Our* galactic disk is drawn out to encompass these cores. The same diagram with all the galactic disks in all six spacetime manifolds drawn out looks as follows:

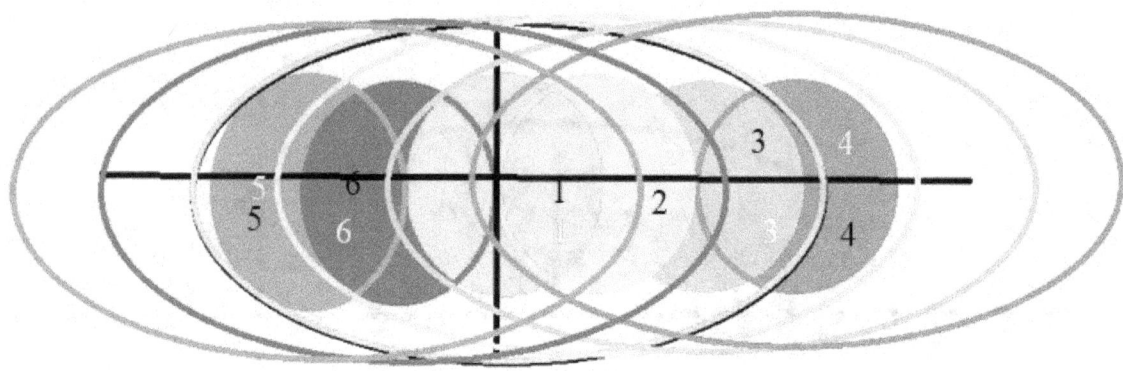

 The circles represent 30,000 light year diameter cores in each spacetime manifold. The ellipses represent the galactic disks in each spacetime manifold. If you draw a sphere from the central point of our galaxy, most of the visible mass is contained in a sphere of radius about 30,000 light years.

You can see from these straightforward assumptions of symmetric but displaced distribution, that a sphere of 15,000 light years from the center of the galaxy contains all the visible mass in the core 30,000 light year diameter of our galaxy and part of the core 30,000 light year diameter mass from spacetime manifolds 2 and 6.

Recall that their impact on mass is of a lower order than the visible mass. Note that the densest parts of each galaxy are shown in the diagram, but all the galaxies are spread out in each spacetime manifold, just like ours is. Our galactic mass will contain 100 billion solar masses in this sphere contributed from the native spacetime manifold, combined with about 320 billion more solar masses from manifolds 2 and 6.

All this mass plus the concentrated core at the center of manifold 2 and 6 (located 15,000 light years from the center of our galaxy), explains the stable orbits of peripheral stars in the galaxy. If you draw a sphere of radius 30,000 light years from the center of the galaxy, you will have the full visible mass in a 60,000 light year diameter of our galaxy, plus the core 30,000 light year diameter contribution from manifolds 2 and 6.

Additionally, you will have partial contributions from the 30,000 light year core diameters in spacetime manifold of 3 and 5. If you draw a sphere of radius 45,000 light years, you have the full 30,000 light year diameter core mass contribution from spacetime manifolds 2, 3 5 and 6. In addition, part of the mass of the 30,000 light year core in manifold 4 is also included.

To include the full mass of all spacetime manifolds, you must draw a sphere of radius 60,000 light years, in which case you include the mass of all six spacetime manifolds in total. All the net halo effects at the edges of the galaxy and towards the Andromeda galaxy are explained by this. The motion of any stars in the periphery is determined by the total mass in a sphere of radius r, where r is the distance of the star from the galactic center.

The total mass in the largest sphere, accounting for all mass, will be about 900 billion solar masses, including all manifolds. This is assuming a base figure of about 160 billion solar masses per spacetime manifolds other than our own. The 900 billion solar masses are crossing manifolds, so the effective mass will be reduced.

Note also that there is some equilibrium in terms of the mass; in current models, the star density drops off mysteriously after a radius of 40,000 light years; in the proposed model, you get a uniform mass distribution, if you view the mass as the sum of all masses in all spacetime manifolds.

For our sun, we have the following approximate calculations:

$v= 220 \ km/sec$ [8.13]

Converting this to parsecs per year using the conversion of 1 km = 3.2408e-14 parsecs and 1 year= 31,556,909 seconds, we get

$v= .000225 \ parsecs/year$ [8.14]

With the radius of a sphere encompassing the sun being $a=8000$ parsecs , we get the circumference of the sun's orbit as:

$C = 2\pi a = 50,300 \ parsecs.$ [8.15]

The period of the orbit is going to be:

$$P = C/v$$

$$= 50{,}300 \text{ parsecs}/.000225 \text{ parsecs/year}$$

$$= 220{,}000{,}000 \text{ years.} \qquad\qquad [8.16]$$

We can do similar calculations for the period of the stars at the outer edge of the galaxy in the six-manifold model. Let us consider a sample radius of about 20,000 parsecs from the center of the galaxy. This is about 20,000 * 206265 AUs from the center of the galaxy, which is about 4.125 billion AUs.

The mass enclosed by a sphere of this radius will encompass all the 'hidden' mass from all the manifolds- approximately 900 billion solar masses, muted by intermanifold effects. We can use Kepler's law now to figure out the period of the stars in this orbit around the center of the galaxy.

$$P^2 = a^3/M \qquad\qquad [8.17]$$

For now, let us assume that the total mass acts as if it is all concentrated in our manifold, as a first approximation. Substituting our estimated values in we get

$$P_{outerstar} = 279 \text{ million years} \qquad\qquad [8.18]$$

Note that this falls in the accepted range of the orbital period of the spiral arm area, usually estimated to be in between 220 million years and 360 million years. This calculation falls almost right in the middle of this range.

Now let us assume some "drag" due to the intermanifold effects. Recall the intermanifold factor that varies between 0 and 1 in limits that represents the effect of one manifold on another. This has quite a bit of variation locally at quantum scales, but it is reasonable to assume that there is an average factor that prevails over galactic scales, which we label **imf**. Note that average values of **imf** are substantially different for galactic scales as opposed to cosmological scales. This is due to the concentration of masses in galaxies in contrast to the void that constitutes most of the universe.

If we assume it averages out to 0.5 over galactic scales, the effective solar masses contained by a sphere of radius 20,000 parsecs will be multiplied by .5, and therefore will be around 450 billion solar masses. In this case, we get:

$P_{outerstar}$ = 394 million years (imf=0.5)   [8.19]

This is slightly larger than the outer estimate of the 220 million years to 360 million years. If we assume it averages out to .75, the effective mass will be .75 times 900 billion solar masses resulting in :

$P_{outerstar}$ = 322 million years (imf=0.75)   [8.20]

The intermanifold factor has variance with various portions of space. It is the dense concentration of masses packed in tight, like supermassive black hole cores, which make this factor vary between .5 and .75 for galaxies. The local variances would help explain some of the eccentricities of the orbits as you go further out.

The following chart shows the projected period in my model, assuming varying factors for imf between .5 and .9 in increments of .25:

Intermanifold factor	Period of outer stars (millions of years)
.5	394
.525	385
.55	377
.575	368
.6	361
.625	353
.65	346
.675	340
.7	334
.725	328
.75	322
.775	317
.8	312
.825	307
.85	303

.875	299
.9	294
.925	290
.975	283
1.0 (never achieved)	279

Our chart matches up quite well of the range of the estimated period for the spiral arm: between 220 million years and 360 million years. Any value of imf between .6 and .975 fits exactly with the observed period. With the amount of concentrated mass in our galaxy (including a supermassive black hole at the center), this is a completely plausible range for the intermanifold factor.

An interesting point here is that the halo effect will be asymmetric according to this model. This should be a measurable effect and is true in any given spacetime manifold.   Note that the eccentric and irregular orbits of the outer stars is a qualitatively natural consequence of the distribution we have suggested; there are competitive attractive forces from different spacetime manifolds contribute to the sphere with the mass affecting the orbit.

To a close approximation, the period of the orbiting star is dependent on the mass enclosed by a sphere out to a radius of the star.

Finally, note that with the proposed reasonable distribution of masses in the six spacetime manifolds, a galactic disk in any spacetime manifold is impacted by at least 3 core galactic masses and a lot of other peripheral mass as well, suggesting that the halo effect would occur in every spacetime manifold (albeit to differing degrees).

We can replace MACHOS, WIMPS and neutrinos with this amazingly simple model! This seems like such a straightforward explanation with some very rudimentary assumptions if you accept the basic validity of the multiple spacetime manifold model. The math required to explain this is at high school level and is very easily understood.

The only assumptions required are:

1) Mass is spread out uniformly over six spacetime manifolds because of the shearing of the 24-dimensional universe into six distinct 4-dimensional spacetime manifolds.
2) Each of the six distinct spacetime manifolds is separated by an intermanifold energy layer. The effects of mass in one spacetime manifold on another is expressed by an intermanifold factor that varies between 0 and 1 since they lie in distinct spacetime manifolds separated by a layer of energy.

N. Dark Matter is Invisible

In our "local" spacetime manifold, all matter is visible to us. Matter in this model is formed by quantum particles that are constantly transitioning between six different spacetime manifolds but form enough cohesion at macroscopic scales to be confined to what appears to be a continuous four-dimensional spacetime continuum.

The stitching of the universe is so fine (meaning that quantization is at such a small scale) that as you zoom out to any appreciable scale all the back-and-forth transitions between the spacetime manifolds are unobservable. At the macroscopic scale, particles form enough cohesion to appear as mass.

Each observer in the different spacetime manifolds sees this as a symmetric proposition. If you happened to be in spacetime manifold 2, all the matter in spacetime manifold 2 is visible to you. However, all matter from another spacetime manifold is invisible to you. This is a natural consequence of the geometry of the universe in this model. The diagram below shows a concentrated source of photons in manifold 1, scattering everywhere in large volumes and striking the intermanifold energy barrier.

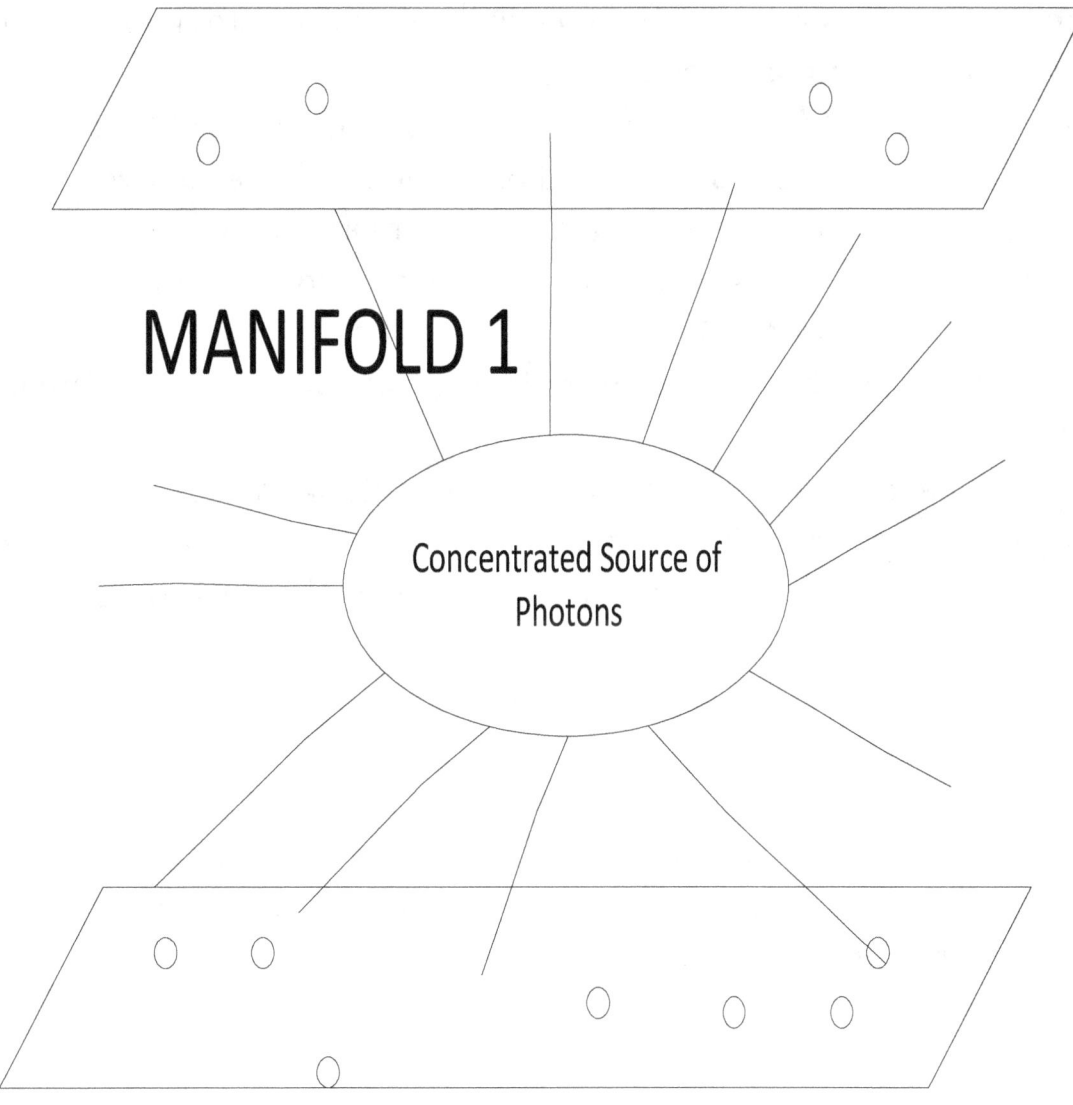

MANIFOLD 6
5% visible mass local to manifold

INTERMANIFOLD ENERGY

MANIFOLD 1

Concentrated Source of Photons

INTERMANIFOLD ENERGY

MANIFOLD 2

Since each manifold is separated by a layer of energy density, a huge volume of photons (such as that from a star) reaching the layer will not penetrate it. This is because they will have to make their way through a quantum fold and past the barrier.

The photon must first encounter a quantum fold in manifold 1 that gets it to either manifold 2 or manifold 6. They may be swallowed up by the barrier to contribute to the pool or on a smaller scale transition back and forth between the manifolds (as will be explained in the renormalization section).

On a large scale they do not make it. They are visible only in the local manifold.

Matter in each spacetime manifold moves along the curves imposed on it by the geometry- it is just that the geometry is 24 dimensional at microscopic scales and merely four dimensional at macroscopic scales. Matter and photons in other spacetime manifolds follow the same laws of physics as ones in ours do.

They do not mysteriously tunnel across manifolds and appear in another one under normal circumstances- only in quantum circumstances, where they reside in the pool of energy density between manifolds.

This is a much more attractive explanation than imbuing dark matter with some sort of mystical properties where it is all around us but does not interact with us for any apparent reason. There is a physical barrier in the form of the intermanifold energy that prevents the interaction.

O. Dark Matter is transparent.

Think of the different manifolds as parallel tram tracks. Everything on each track moves in a precise defined way but is confined to the local track for the most part. The photons that move along local tracks stay there for the most part, kept apart by intermanifold energy and therefore have no material impact on the way we visibly measure the universe.

The diagram below depicts the situation of a star in our manifold trying to send its photons across the manifolds:

MANIFOLD 6
5% visible mass local to manifold

INTERMANIFOLD ENERGY

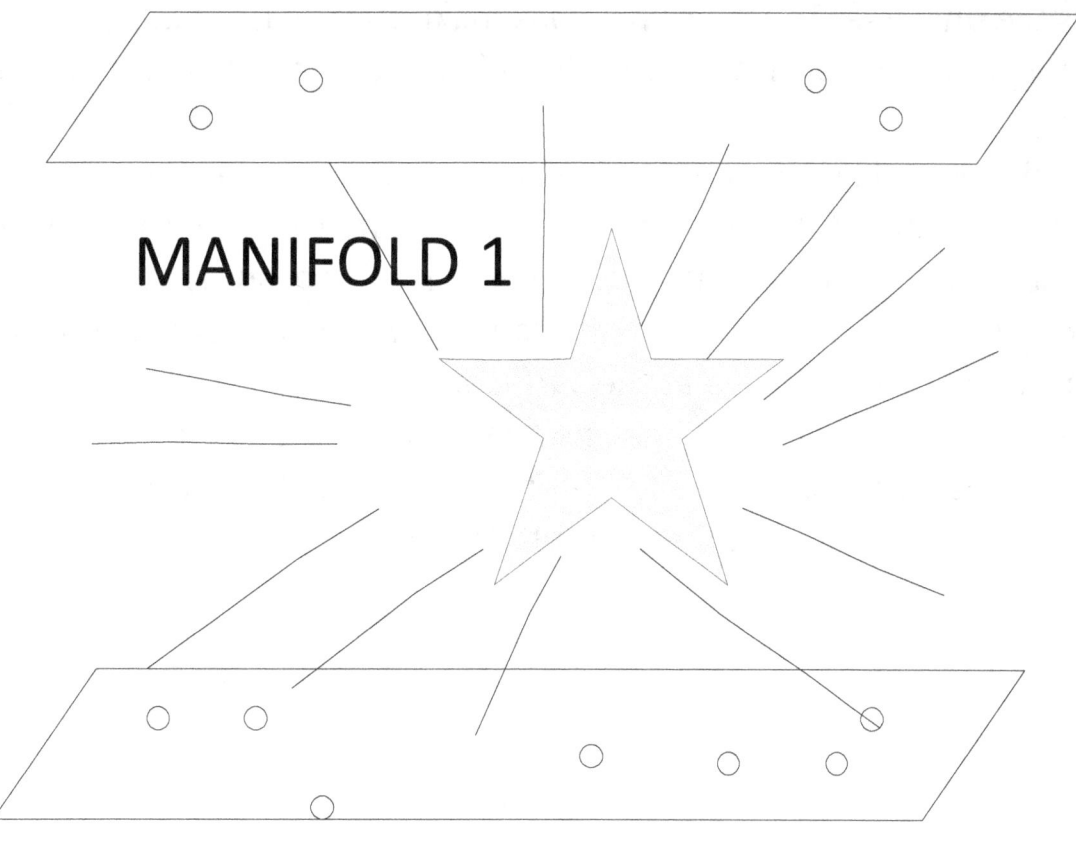

MANIFOLD 1

INTERMANIFOLD ENERGY

MANIFOLD 2

The photons from the star are inhibited from traveling to adjacent manifolds by both the size of the quantum folds (microscopic in nature) and the traversal across the intermanifold energy barrier.

Whatever photons attempt this treacherous long journey are swallowed up by the intermanifold energy pool. The few that survive will affect only some quantum mechanical processes and not any macroscopic views.

All the photons that stay local to the native manifold constitute the visible universe. The other photons making transitions in and out of the energy density layer have a profound effect on the quantum universe only.

If this seems like a confusing explanation think about this: a photon from the core of the sun may take 5000 years to get to the surface.

What we think of as continuous emission of photons is actually a long process due to the collisions and random walks. It takes several steps for the photon to arrive at the surface!

That being the case, just imagine traversing across an intermanifold energy layer with a soup of photons and elementary particles.

It is small wonder that the poor photons never make it.

P. Dark Matter interacts gravitationally with matter in our universe even though it does not in other ways.

This again is due to the geometry of the universe. All particles that might mediate other forces interact with the intermanifold energy density layer. They contribute to the pool of energy that makes up the layer or they slip through quantum folds and have an impact at microscopic scales only.

Each manifold is surrounded by its opposite polarity matter. For example, manifold 1 is surrounded by manifold 6 and manifold 2 both of which are antimatter dominant. The energy layer between these manifolds is a soup of elementary particles and their antimatter counterparts annihilating and reforming.

This creates what we perceive as the vacuum energy in manifold 1. This prevents macroscopic matter to interact via electromagnetic, weak or strong forces. The photons, gluons, W-bosons, and Z-bosons simply have no chance of penetrating the intermanifold energy barrier to make an interaction possible on any mass scale. To the extent there is any interaction, it is at ultramicroscopic scales in the quantum fold.

The following diagram shows a sample mass in manifold 1 and another one in manifold 2:

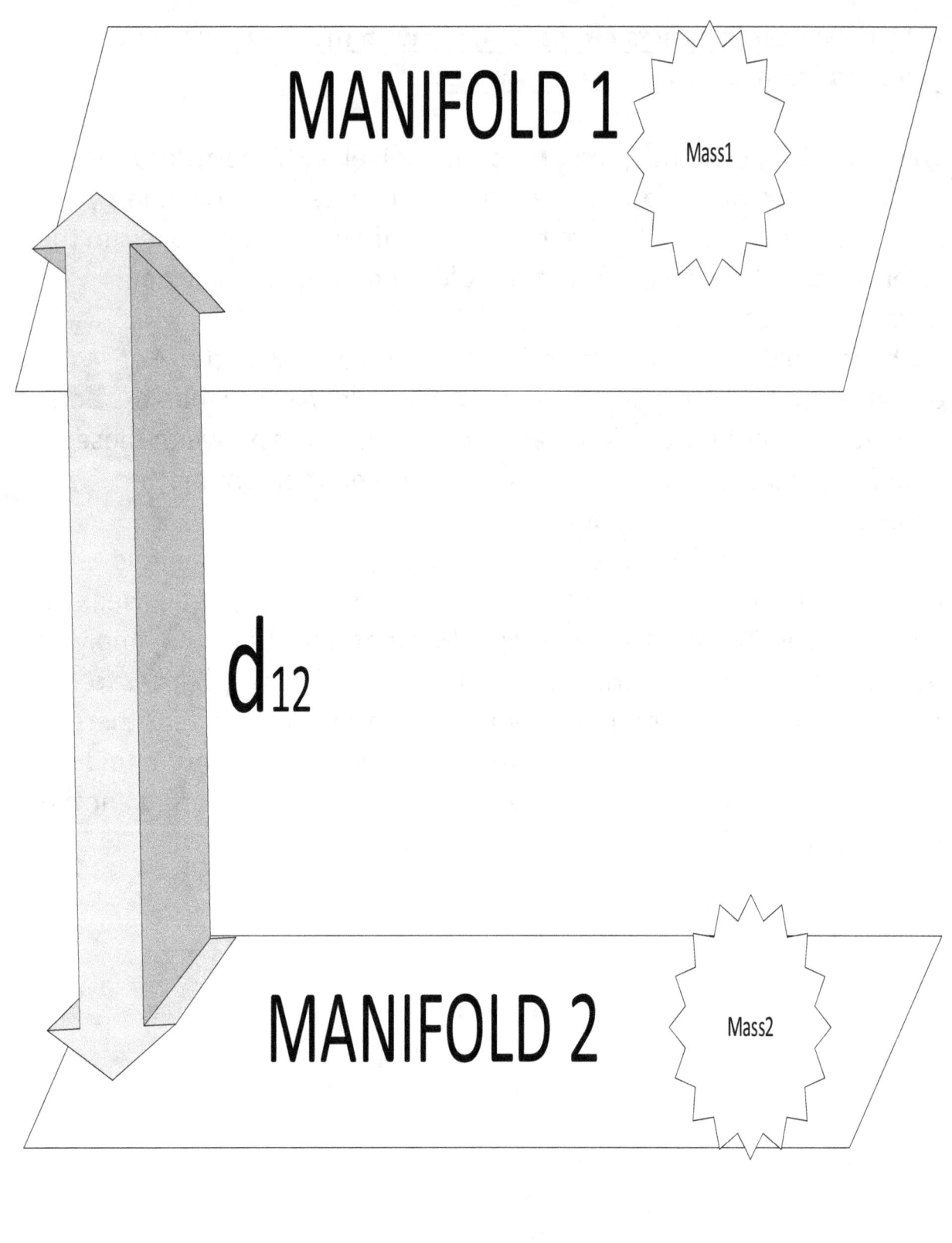

The masses are separated by an intermanifold energy layer and a factor d_{12} representing the intermanifold separation. Normally, the interaction would be proportional to the masses multiplied by the inverse square of the distance.

However, as we have established as an assumption, the intermanifold energy layer between these manifolds is constantly pressuring them apart. It has been that way since the Big Bang. Time moving forward corresponds directly to the explosion pushing these six manifolds apart. Reversing time lets you go backwards to the originating singularity that gave birth to the universe.

Thus, gravity will have an impact even across the barrier, but attenuated by the intermanifold energy layer and separation between the manifolds.

This does not mean the effect is always negligible. With the interweaving of the six-spacetime manifolds, a massive object stretches the manifold in one spacetime, it stretches so far that the impact on another one is significant.

Extending our metaphor of parallel trams, when a car on the tram is so heavy that the entire track deforms downward to nearly touch the track below it, both tracks are affected.

These extremely massive cases are where quantum mechanics and gravity meet. In my model, this is reflected by the increase of the effect of gravity. This increase is identified with dark matter in conventional models but is merely ordinary mass in both manifolds in my model. Accompanying the increased gravitational effects will also be a stretching of the quantum folds and a marked increase in quantum activity between manifolds.

Q. Dark Matter forms an asymmetric amount of matter in the universe.

Not in the proposed model! All the matter in the six separate manifolds is symmetrically distributed. Each observer classifies all extra manifold matter as 'dark' and hence the seeming asymmetry. This is depicted in the diagram below:

MANIFOLD 1
5% visible mass local to manifold

MANIFOLD 2
5% visible mass local to manifold

MANIFOLD 3
5% visible mass local to manifold

MANIFOLD 4
5% visible mass local to manifold

MANIFOLD 5
5% visible mass local to manifold

MANIFOLD 6
5% visible mass local to manifold

Each manifold views the 25% extra-manifold matter as "dark matter". The intermanifold energy layers between the manifolds comprises what is labeled as "dark energy" in conventional models (11.67% per layer making up 70% of the universe). This adds some beautiful symmetry to the universe and eliminates the need for an ugly inexplicable asymmetry.

R. Dark Matter causes macro and micro gravitational lensing.

 As dense massive objects stretch the local spacetime manifold, the deformation indirectly affects all spacetime manifolds. Imagine two rubber sheets on top of each other with a massive indentation in the first one even when the second one is flat. Even if the massive indentation does not touch the flat one, it will affect the flat sheet (gravitationally).

Dense large objects in one manifold would cause macroscopic gravitational lensing in another. Dense small objects such as mini black holes in one would cause microscopic gravitational lensing in another.

The dense large mass case between manifold 1 and 2 is depicted below:

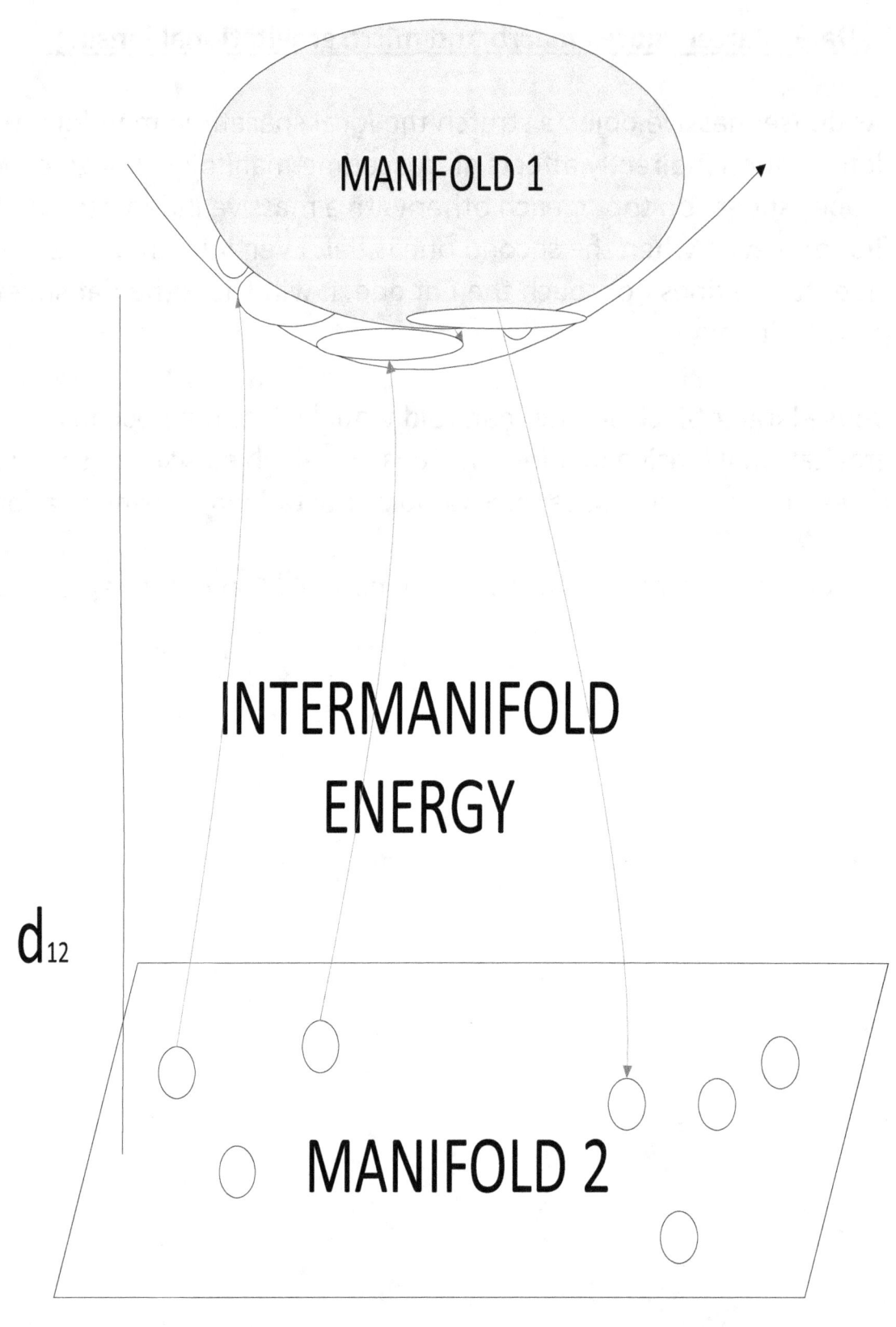

Two contributing factors cause the gravitational lensing. Unlike the ordinary case where photons are doing intermittent transient interactions between narrow quantum folds, the folds get stretched by the dense large mass.

The very fabric of spacetime is getting stretched, distorting the size of the quantum folds.

Therefore, it is easier for the photons to get through the intermanifold energy layer, and once through they are subject to the same General Relativity laws as in any single spacetime manifold.

Note that the mass must be extremely large and extremely dense for this case to occur. Ordinarily photons are not slipping through easily (otherwise, dark matter would be visible).

The situation below depicts a small dense mass and its effect on manifold 1 and 2.

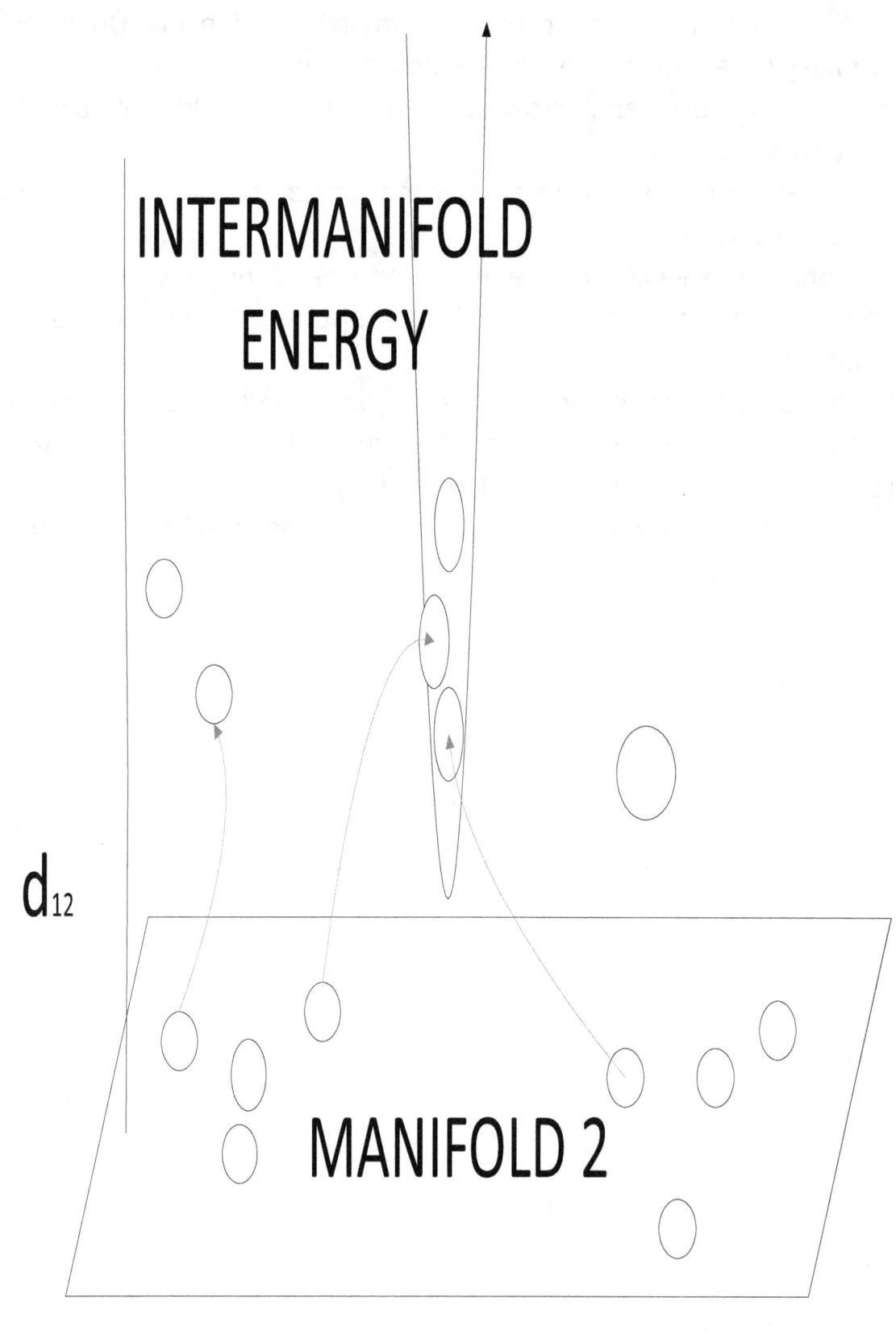

In this situation, the protrusion into manifold 2 from manifold 1 is a narrow but deep one. The narrow but deep feature causes several quantum folds to stretch, albeit not as much as in a dense large mass. The principle is the same, it is just over a narrower area. Photons flood in from the intermanifold region and suffer micro gravitational lensing as a result.

S. Different areas of the universe are affected differently by dark matter.

The influence of dark matter in the universe is not uniform. In some regions, it accounts for larger percentage of the total mass. This is very naturally explained in my proposed model because the influence depends on the clumps of matter present in your local area.

This quantity averages out to a certain amount over the entire cosmos, but its density varies locally. For a given region of space, the intermanifold separation can be high and/or there might be a high concentration of extra-manifold mass (which is tagged as dark matter in conventional theories).

For example, our Sun may be proximal to another star in an adjacent manifold, resulting in some pronounced effects. On the other hand, another star in our manifold may be proximal to an interstellar void in an adjacent manifold, yielding minimal effects. This explains some gravitational anomalies in the galaxy, where it seems like unseen masses are affecting motion.

The following scenarios should help in understanding the varying effects. In the following scenario, a local patch of 3 stars in manifold 1 is adjacent to a void in manifold 2:

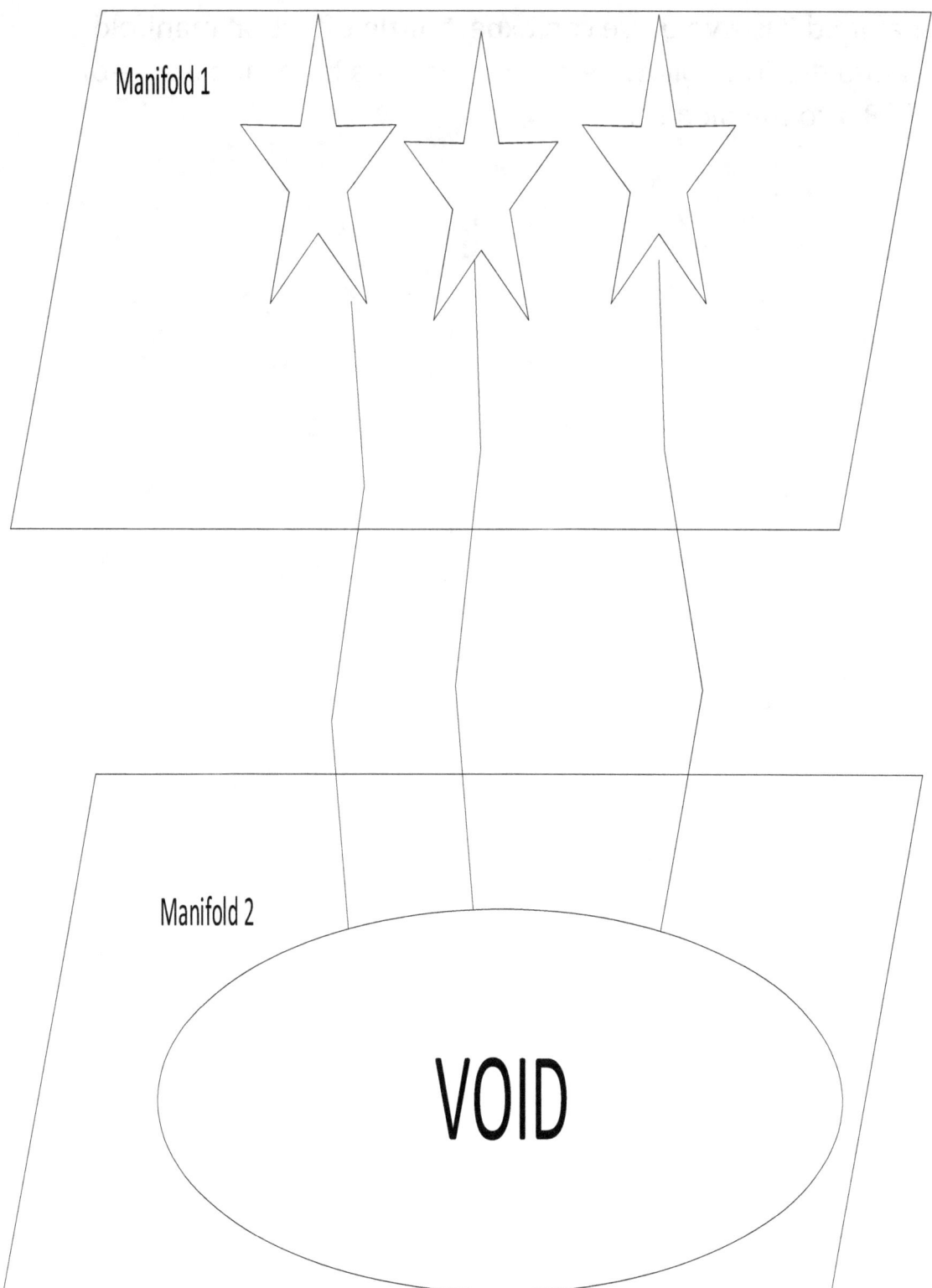

Since manifold 2 is a void, we can expect little effect on manifold 1 from manifold 2. The following diagram factors in the local area of manifold 3 into the picture:

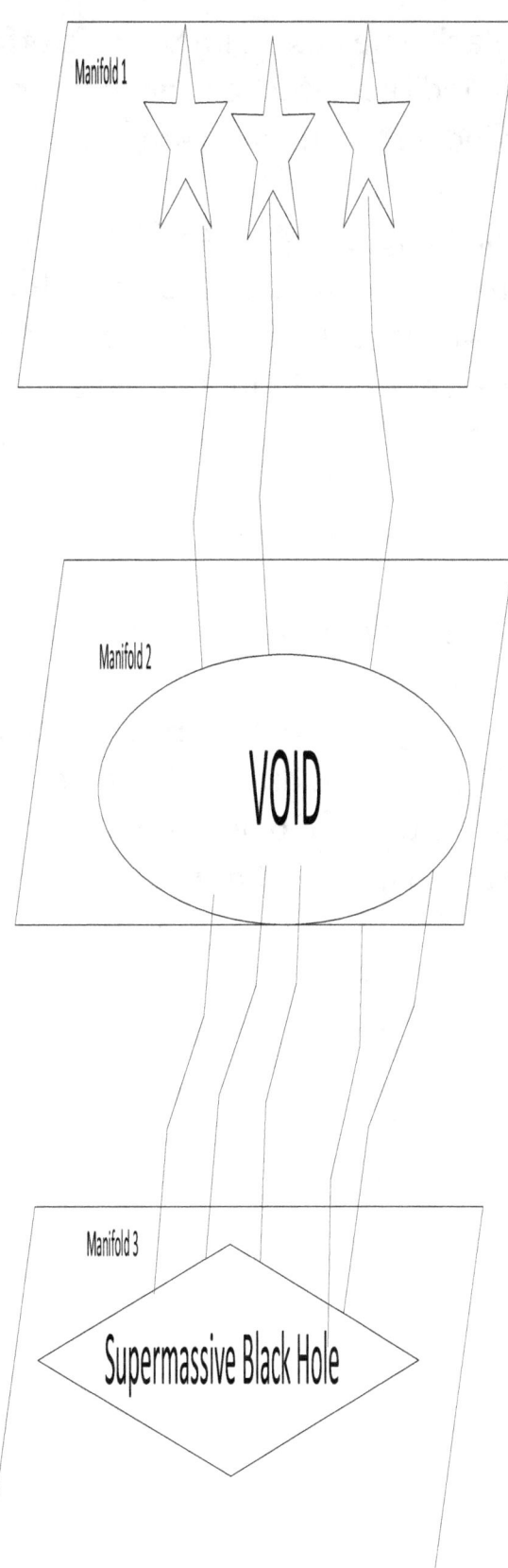

The supermassive black hole has an immediate effect even on the void in manifold 2 and an indirect effect on the local star cluster in manifold 1. We can keep extending this to all the manifolds.

At the time of the Big Bang, the six manifolds were cleaved off and the distribution of the mass is random in each one. It is obviously likely that different areas will have different concentrations of mass relative to each other. Moreover, the situation is dynamic, as there are rotations around galactic cores and relative movement between galaxies in the universe. It is little wonder that the effects of the extramanifold mass are different for different areas of the universe.

An analogy that will help is to think of the six different manifolds as six pool tables stacked on top of each other. You open with a break on each table, which disperses the balls across the tables.
If you take a ball on table 1 and draw a line down towards table 2, you may hit an area with a cluster of balls, or you might just hit empty space, or you might hit an area with another ball.
If you further assume the different numbered balls represent different density and concentration of masses, it will provide a clear picture of how an adjacent manifold can locally vary in impact to the original manifold.
As the game progresses and the players start sinking balls in various pockets the effects vary dynamically, like the factoring of dynamic movement of stars , galaxies and galactic clusters.

T. Special Relativity

We must reconcile our claims of a 24-dimensional superstructure with what we see as four-dimensional spacetime in our manifold. The invariant intervals local to each spacetime manifold in terms of the manifold-specific coordinates are as follows:

$$ds_1^2 = dx_0^2 - dx_1^2 - dx_2^2 - dx_3^2 \text{ (spacetime manifold 1)} \qquad [8.21]$$
$$ds_2^2 = dx_4^2 - dx_5^2 - dx_6^2 - dx_7^2 \text{ (spacetime manifold 2)}$$
$$[8.22]$$
$$ds_3^2 = dx_8^2 - dx_9^2 - dx_{10}^2 - dx_{11}^2 \text{ (spacetime manifold 3)}$$
$$[8.23]$$
$$ds_4^2 = dx_{12}^2 - dx_{13}^2 - dx_{14}^2 - dx_{15}^2 \text{ (spacetime manifold 4)}$$
$$[8.24]$$
$$ds_5^2 = dx_{16}^2 - dx_{17}^2 - dx_{18}^2 - dx_{19}^2 \text{ (spacetime manifold 5)}$$
$$[8.25]$$
$$ds_6^2 = dx_{20}^2 - dx_{21}^2 - dx_{22}^2 - dx_{23}^2 \text{ (spacetime manifold 6)}$$
$$[8.26]$$

Note that here $dx_0^2 = c^2 (dt_0)^2$, $dx_4^2 = c^2 (dt_1)^2$, $dx_8^2 = c^2 (dt_2)^2$, $dx_{12}^2 = c^2 (dt_3)^2$, $dx_{16}^2 = c^2 (dt_4)^2$ and $dx_{20}^2 = c^2 (dt_5)^2$.

The generalized invariant, factoring in the intermanifold factor is :

$$ds_{11}^2 = ds_1^2 + (imf_{12}ds_2)^2 + (imf_{13}ds_3)^2 + (imf_{14}ds_4)^2 + (imf_{15}ds_5)^2 + (imf_{16}ds_6)^2 \qquad [8.27]$$

Recall that the size of the quantum fold squared is given by the following: $(imf_{12}ds_2)^2 + (imf_{13}ds_3)^2 + (imf_{14}ds_4)^2 + (imf_{15}ds_5)^2 + (imf_{16}ds_6)^2$ =**magnitude of ($\hbar^2$ /2).** Therefore, on any scales sufficiently greater than the quantum fold, this becomes equivalent to exactly what we expect. If the scale of the object and the motion is larger than a number of quantum folds, then the last five terms in the equation become irrelevant. Thus, my model reduces to "regular" special relativity for all practical cases.

The generalized invariant interval captures the structure of the universe in my proposed model. Each spacetime manifold is indirectly affected by the other ones, but the effect depends on the separation between the manifolds, populated by intermanifold energy layers.

Similarly, we have the following results for the other 5 spacetime manifolds.

$$ds_{ii}^2 = ds_i^2 + \sum_{l \neq j} (Imf_{ij}ds_j)^2 \qquad\qquad [8.28]$$

where the index i varies from 1 to 6 and the summation prohibits the condition i=j and j also run from 1 to 6. For almost all situations, the generalized invariant interval is going to be equivalent to the standard one. Only where intermanifold distances are exceedingly small will there be modifications and perturbations to it. The metric ds_{ii}^2 (i=1,2,3,4,5,6) will be the basic definition of invariant intervals in the universe made of six spacetime manifolds. We can define a generalized Lorentz transformation for our spacetime manifold (manifold 1) as:

$$x^{\mu'} = \Lambda^{\mu}_{\nu} x^{\nu}$$

$$[8.29]$$

$\Lambda^{\mu}{}_{\upsilon}$ is a 24 by 24 matrix, with component elements. It is easier to write this is 6 4 by 4 blocks for readability and to distinguish the separate spacetime manifolds easily:

$$\Lambda^{\mu}_{\nu} = \begin{pmatrix} (\Lambda_1)^{\mu}_{\nu} & 0 & 0 & 0 & 0 & 0 \\ 0 & (\Lambda_2)^{\mu}_{\nu} & 0 & 0 & 0 & 0 \\ 0 & 0 & (\Lambda_3)^{\mu}_{\nu} & 0 & 0 & 0 \\ 0 & 0 & 0 & (\Lambda_4)^{\mu}_{\nu} & 0 & 0 \\ 0 & 0 & 0 & 0 & (\Lambda_5)^{\mu}_{\nu} & 0 \\ 0 & 0 & 0 & 0 & 0 & (\Lambda_6)^{\mu}_{\nu} \end{pmatrix}$$

[8.30]

where

$$(\Lambda_1)^{\mu}_{\nu} = \begin{pmatrix} \gamma & -\gamma v/c & 0 & 0 \\ -\gamma v/c & \gamma & 0 & 0 \\ 0 & 0 & 1 & 0 \\ 0 & 0 & 0 & 1 \end{pmatrix}$$

[8.31]

and

$$\gamma = \frac{1}{\sqrt{1 - \dfrac{v^2}{c^2}}}$$

[8.32]

This is just the usual Lorentz transformation that plays a role in time dilation and length contraction. For the other manifolds, the intermanifold factor must modulate this so that it plays virtually no role in normal circumstances but starts affecting events when there is crossover between the manifolds. This is very simply accomplished with the following definitions:

$$(\Lambda_2)^\mu_\nu = imf_{12}(\Lambda_1)^\mu_\nu$$

[8.33]

$$(\Lambda_3)^\mu_\nu = imf_{13}(\Lambda_1)^\mu_\nu$$

[8.34]

$$(\Lambda_4)^\mu_\nu = imf_{14}(\Lambda_1)^\mu_\nu$$

[8.35]

$$(\Lambda_5)^\mu_\nu = imf_{15}(\Lambda_1)^\mu_\nu$$

[8.36]

$$(\Lambda_6)^\mu_\nu = imf_{16}(\Lambda_1)^\mu_\nu$$

[8.37]

Recall that *imf* is a factor that is remarkably close to zero under conditions where the intermanifold distance is large. Thus, this matrix, which does look overly complex relative to the usual Lorentz transformation, simplifies to the usual 4 by 4 matrix that we are accustomed to.

On the other hand, the other extreme case would be where all the manifolds are in proximity, you get a Lorentz transformation per manifold. They are nearly symmetric transformations in every other manifold (not quite equal, because the intermanifold factor never quite reaches 1, it is just very nearly 1 for small values of intermanifold separation). The Minkowski metric in our spacetime manifold looks like:

$$(\eta_1)^{\mu}_{\nu} = \begin{pmatrix} 1 & 0 & 0 & 0 \\ 0 & -1 & 0 & 0 \\ 0 & 0 & -1 & 0 \\ 0 & 0 & 0 & -1 \end{pmatrix}$$

[8.38]

The generalized Minkowski metric must include this as a subset and is defined as:

$$
\eta^{\mu}_{\nu} = \begin{pmatrix}
(\eta_1)^{\mu}_{\nu} & 0 & 0 & 0 & 0 & 0 \\
0 & (\eta_2)^{\mu}_{\nu} & 0 & 0 & 0 & 0 \\
0 & 0 & (\eta_3)^{\mu}_{\nu} & 0 & 0 & 0 \\
0 & 0 & 0 & (\eta_4)^{\mu}_{\nu} & 0 & 0 \\
0 & 0 & 0 & 0 & (\eta_5)^{\mu}_{\nu} & 0 \\
0 & 0 & 0 & 0 & 0 & (\eta_6)^{\mu}_{\nu}
\end{pmatrix}
$$

[8.39]

In our spacetime manifold, this is the usual metric (manifold 1), which we designated as $\eta_1{}^{\mu}$

While for the other spacetime manifolds , we have for i=2,3,4,5 and 6.

$$
(\eta_i)^{\mu}_{\nu} = im f_{1i}(\eta_1)^{\mu}_{\nu}
$$

[8.40]

The metric is modulated by the intermanifold factor, which governs the extent to which the manifolds interact in any given region of spacetime in our manifold.

The usual definitions of four velocity, relativistic kinetic energy, total relativistic energy, four momentum and the famous relationship between energy, mass and momentum are all the same – you must apply them to each manifold and apply the intermanifold factor where appropriate. It all comes out of our definition of the generalized metric.

U. General Relativity

Let us take another look at how our universe looks between manifold 1 and manifold 2 (a similar picture holds for all other manifolds).

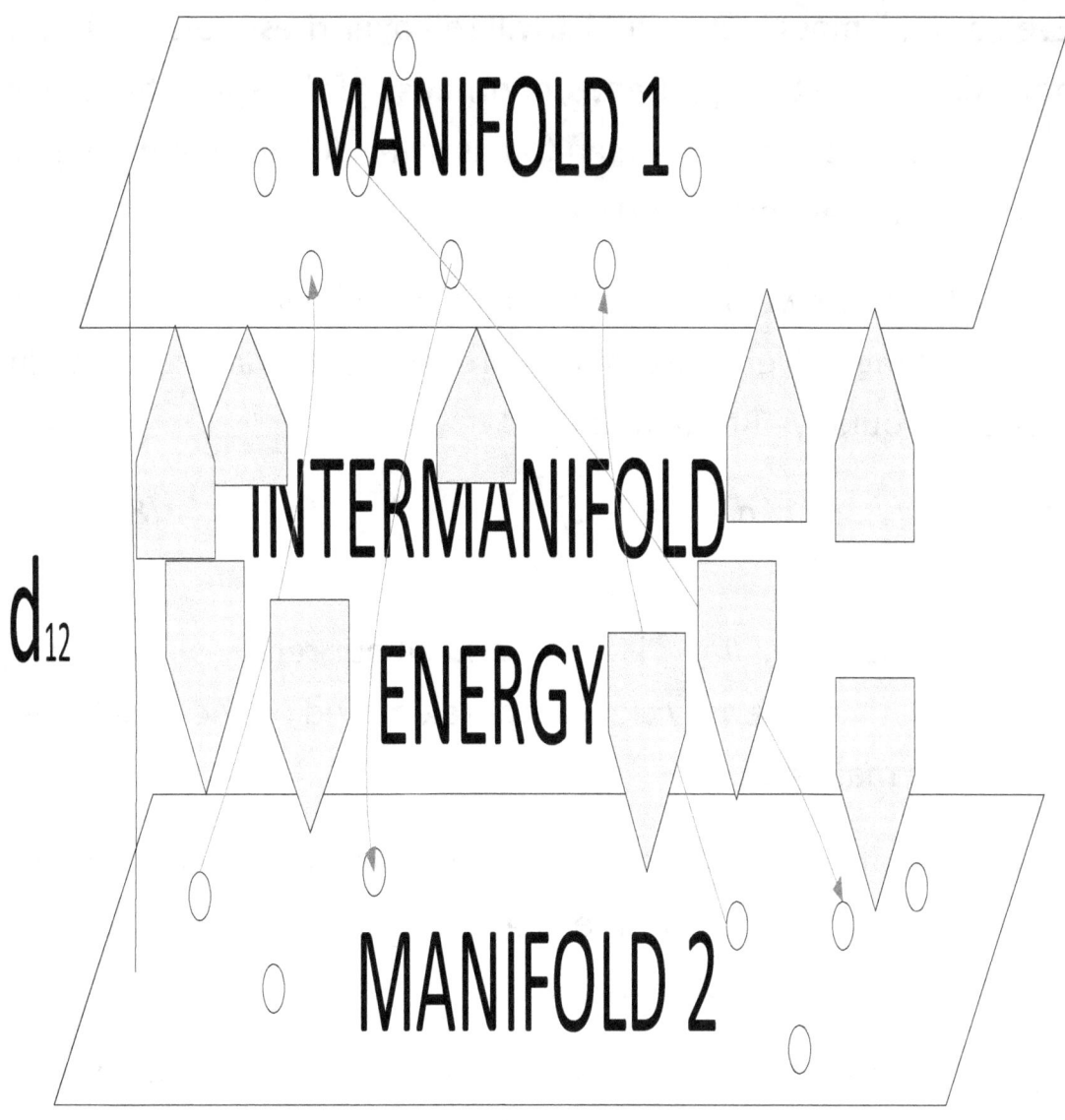

This is in the limit of a flat universe with little mass – not an interesting universe, but it represents the universe in the limit of special relativity.

Those small holes with the red lines are the quantum folds through which small scale particles transition.

Even a small mass in manifold 1 will roll over these folds as if they do not exist, hence the apparent smoothness of the space time manifold. In between the spacetime manifold is layer of intermanifold energy, pushing the manifolds apart.

The distance between the manifolds d_{12} will be large in our low mass uninteresting universe. Nothing is present to distort the spacetime fabric profoundly. Recall that

$$Imf_{12} = d_{12} \left\{ \left[1 + (1/ d_{12}{}^2) \right]^{1/2} - 1 \right\} \qquad\qquad [8.41]$$

Thus, if d_{12} is large, Imf_{12} is small, and all our relativistic equations in special relativity are very closely approximated by the usual ones in our spacetime manifold.

It was part of our fundamental assumption that the Big Bang cleaved these manifolds starting at the beginning. As time moves forward this energy and pressure just increases.

If you roll time backwards all the separate manifolds come together to form the cohesive 24-dimensional Big Bang singularity. The arrow of time is defined by increasing/decreasing intermanifold energy density/pressure between manifolds in my model.

Now let us introduce some massive object into the universe to make it more interesting. This is the province of General Relativity. Look at what happens to our picture.

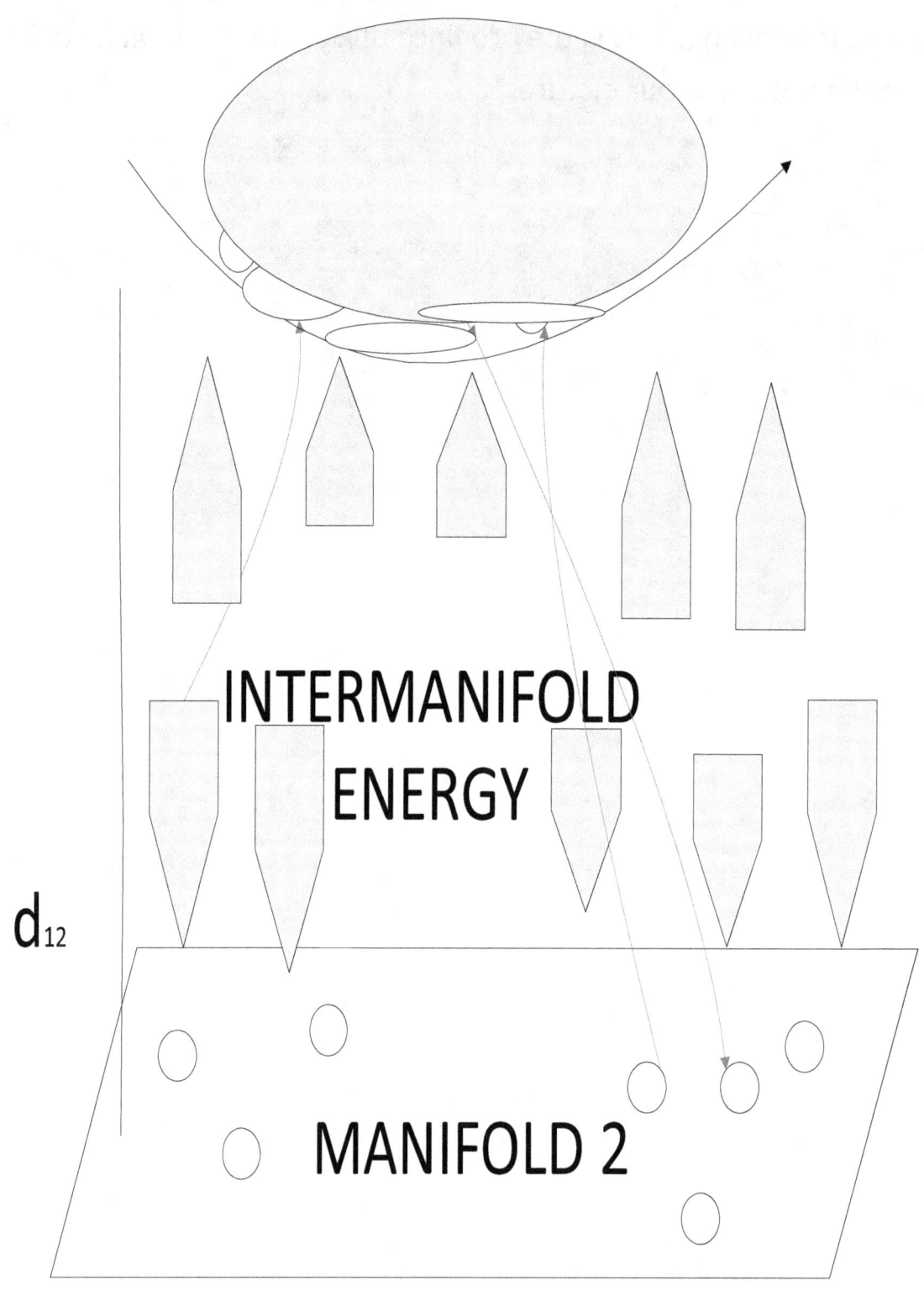

The mass stretches the spacetime fabric and the effects of gravity are counteracted by the effects of the intermanifold energy. This is exactly what "dark energy" is supposed to do – provide a counterbalance to gravity.

So long as the quantum folds do not stretch too much, we will be in the limit of General Relativity. But we must factor in the effects from the other manifolds. Qualitatively, it looks like we should get a repulsive counteractive force.

Let us analyze the Einstein field equations for all six manifolds in this context. These equations are in my mind the most beautiful in theoretical physics. They are obviously complex in terms of solutions, but the vision behind them and the basic philosophy is like an inspirational painting, like the Mona Lisa. The geometry of space is on the left side and a stress energy tensor is on the right side. With no intermanifold effects, the Einstein field equations look as follows, local to each manifold.

If you live in Manifold 1:

$$G_{00} = -\kappa T_{00} \qquad\qquad [8.42]$$
$$G_{01} = -\kappa T_{01} \qquad\qquad [8.43]$$
$$G_{02} = -\kappa T_{02} \qquad\qquad [8.44]$$
$$G_{03} = -\kappa T_{03} \qquad\qquad [8.45]$$

$$G_{11} = -\kappa T_{11} \qquad\qquad [8.46]$$
$$G_{12} = -\kappa T_{12} \qquad\qquad [8.47]$$
$$G_{13} = -\kappa T_{13} \qquad\qquad [8.48]$$

$$G_{22} = -\kappa T_{22} \qquad\qquad [8.49]$$
$$G_{23} = -\kappa T_{23} \qquad\qquad [8.50]$$
$$G_{33} = -\kappa T_{33} \qquad\qquad [8.51]$$

With the usual definitions as follows:

$$G_{\mu\nu} = R_{\mu\nu} - (Rg_{\mu\nu}/2) \qquad\qquad [8.52]$$

$R_{\mu\nu}$ is the Ricci Tensor and R is the Ricci scalar and $g_{\mu\nu}$ is the metric tensor.

Einstein's constant κ is $8\pi G/c^4$. The interpretation of this is loosely described as the relationship between matter (on the right side in the form of a stress energy tensor) and spacetime geometry (as described by the metric tensor and other terms on the left). The nutshell of what happens is that the distinct manifolds all contribute to the behavior of matter on a global scale. Let us see how this shows up in our equations.

If you live in Manifold 2:

$$G_{44} = -\kappa T_{44} \qquad\qquad [8.53]$$
$$G_{45} = -\kappa T_{45} \qquad\qquad [8.54]$$
$$G_{46} = -\kappa T_{46} \qquad\qquad [8.55]$$
$$G_{47} = -\kappa T_{47} \qquad\qquad [8.56]$$

$$G_{55} = -\kappa T_{55} \qquad\qquad [8.57]$$
$$G_{56} = -\kappa T_{56} \qquad\qquad [8.58]$$
$$G_{57} = -\kappa T_{57} \qquad\qquad [8.59]$$

$$G_{66} = -\kappa T_{66} \qquad\qquad [8.60]$$
$$G_{67} = -\kappa T_{67} \qquad\qquad [8.61]$$

$$G_{77} = -\kappa T_{77} \qquad [8.62]$$

If you live in Manifold 3:

$$G_{88} = -\kappa T_{88} \qquad [8.63]$$
$$G_{89} = -\kappa T_{89} \qquad [8.64]$$
$$G_{810} = -\kappa T_{810} \qquad [8.65]$$
$$G_{811} = -\kappa T_{811} \qquad [8.66]$$

$$G_{99} = -\kappa T_{99} \qquad [8.67]$$
$$G_{910} = -\kappa T_{910} \qquad [8.68]$$
$$G_{911} = -\kappa T_{911} \qquad [8.69]$$

$$G_{1010} = -\kappa T_{1010} \qquad [8.70]$$
$$G_{1011} = -\kappa T_{1011} \qquad [8.71]$$
$$G_{1111} = -\kappa T_{1111} \qquad [8.72]$$

If you live in Manifold 4:

$$G_{1212} = -\kappa T_{1212} \qquad [8.73]$$
$$G_{1213} = -\kappa T_{1213} \qquad [8.74]$$
$$G_{1214} = -\kappa T_{1214} \qquad [8.75]$$
$$G_{1215} = -\kappa T_{1215} \qquad [8.76]$$

$$G_{1313} = -\kappa T_{1313} \qquad [8.77]$$
$$G_{1314} = -\kappa T_{1314} \qquad [8.78]$$
$$G_{1315} = -\kappa T_{1315} \qquad [8.79]$$
$$G_{1414} = -\kappa T_{1414} \qquad [8.80]$$
$$G_{1415} = -\kappa T_{1415} \qquad [8.81]$$
$$G_{1515} = -\kappa T_{1515} \qquad [8.82]$$

If you live in Manifold 5:

$$G_{1616} = -\kappa T_{1616} \qquad\qquad [8.83]$$
$$G_{1617} = -\kappa T_{1617} \qquad\qquad [8.84]$$
$$G_{1618} = -\kappa T_{1618} \qquad\qquad [8.85]$$
$$G_{1619} = -\kappa T_{1619} \qquad\qquad [8.86]$$

$$G_{1717} = -\kappa T_{1717} \qquad\qquad [8.87]$$
$$G_{1718} = -\kappa T_{1718} \qquad\qquad [8.88]$$
$$G_{1719} = -\kappa T_{1719} \qquad\qquad [8.89]$$
$$G_{1818} = -\kappa T_{1818} \qquad\qquad [8.90]$$
$$G_{1819} = -\kappa T_{1819} \qquad\qquad [8.91]$$
$$G_{1919} = -\kappa T_{1919} \qquad\qquad [8.92]$$

And finally if you live in Manifold 6:

$$G_{2020} = -\kappa T_{2020} \qquad\qquad [8.93]$$
$$G_{2021} = -\kappa T_{2021} \qquad\qquad [8.94]$$
$$G_{2022} = -\kappa T_{2022} \qquad\qquad [8.95]$$
$$G_{2023} = -\kappa T_{2023} \qquad\qquad [8.96]$$

$$G_{2121} = -\kappa T_{2121} \qquad\qquad [8.97]$$
$$G_{2122} = -\kappa T_{2122} \qquad\qquad [8.98]$$
$$G_{2123} = -\kappa T_{2123} \qquad\qquad [8.99]$$

$$G_{2222} = -\kappa T_{2222} \qquad\qquad [8.100]$$
$$G_{2223} = -\kappa T_{2223} \qquad\qquad [8.101]$$
$$G_{2323} = -\kappa T_{2323} \qquad\qquad [8.102]$$

Although the indices look unwieldy, the basic principle is quite simple. To an exceptionally good first approximation, the Einstein field equations are valid within each spacetime manifold. Einstein later added a cosmological constant term to keep a static universe, and then realized the universe was expanding, and scrapped it. He regarded the term as a blunder.

Many recent developments have shown that this blunder is a valid term, although the rationale for it is unclear. When you are a genius like Einstein, even your apparent blunders turn out to be a true representation of reality. I will demonstrate shortly that it is the perturbations from other manifolds that give rise to the cosmological constant very naturally.

Let us focus in on manifold 1, which we are assuming is our native manifold, and treat all the other ones as perturbations:

$$R_{00}-(Rg_{00}/2)= -\kappa T_{00} \tag{8.103}$$
$$R_{01}-(Rg_{01}/2)= -\kappa T_{01} \tag{8.104}$$
$$R_{02}-(Rg_{02}/2)= -\kappa T_{02} \tag{8.105}$$
$$R_{03}-(Rg_{03}/2)= -\kappa T_{03} \tag{8.106}$$
$$R_{11}-(Rg_{11}/2)= -\kappa T_{11} \tag{8.107}$$
$$R_{12}-(Rg_{12}/2)= -\kappa T_{12} \tag{8.108}$$
$$R_{13}-(Rg_{13}/2)= -\kappa T_{13} \tag{8.109}$$
$$R_{22}-(Rg_{22}/2)= -\kappa T_{22} \tag{8.110}$$
$$R_{23}-(Rg_{23}/2)= -\kappa T_{23} \tag{8.111}$$
$$R_{33}-(Rg_{33}/2)= -\kappa T_{33} \tag{8.112}$$

The counterparts to R_{00} in equation 8.103 for the other manifolds are $R_{44}, R_{88}, R_{1212}, R_{1616}, R_{2020}$. The fundamental effect on g_{00} by the perturbation introduced by the other manifolds is as follows: it is modified by the intermanifold factor imf_{12} for manifold 2, imf_{13} for manifold 3, imf_{14} for manifold 4, imf_{15} for manifold 5 and imf_{16} for manifold 6. The Ricci scalar R involves a contraction of the Ricci tensor and sums over terms like g_{44}; but this term itself involves a multiple of imf_{12} and g_{00}. Consequently, the left side of the equation 8.86 for manifold 1 looks like:

$$R_{00} + R_{44} + R_{88} + R_{1212} + R_{1616} + R_{2020} + R_{2323} - (Rg_{00}/2) - ([imf_{12}]^2 Rg_{00}/2) - ([imf_{13}]^2 Rg_{00}/2) - ([imf_{14}]^2 Rg_{00}/2) - ([imf_{15}]^2 Rg_{00}/2) - ([imf_{16}]^2 Rg_{00}/2)$$
$$[8.113]$$

Note that since the intermanifold factor is less than 1 the squared term is smaller subtraction than the native manifold term. On the right side, we have modified stress energy tensor from each manifold contributing to the total:

$$-\kappa T_{00} - \kappa imf_{12} T_{44} - \kappa imf_{13} T_{88} - \kappa imf_{14} T_{1212} - \kappa imf_{15} T_{1616}$$
$$-\kappa imf_{16} T_{2020} \qquad\qquad [8.114]$$

Gathering up all the terms, we have:

$$R_{00} - (Rg_{00}/2) + \{R_{44} + R_{88} + R_{1212} + R_{1616} + R_{2020} + R_{2323}\} - ([imf_{12}]^2 Rg_{00}/2)$$
$$- ([imf_{13}]^2 Rg_{00}/2) - ([imf_{14}]^2 Rg_{00}/2) - ([imf_{15}]^2 Rg_{00}/2) - ([imf_{16}]^2 Rg_{00}/2)$$
$$+ \kappa imf_{12} T_{44} + \kappa imf_{13} T_{88} + \kappa imf_{14} T_{1212} + \kappa imf_{15} T_{1616} + \kappa imf_{16} T_{2020} = -\kappa T_{00}$$
$$[8.115]$$

Let us make the following identifications:

$$R_{00}' = R_{00} + \{R_{44} + R_{88} + R_{1212} + R_{1616} + R_{2020} + R_{2323}\}$$

$$Rg_{00}/2' = \{1 + [imf_{12}]^2 + [imf_{13}]^2 + [imf_{14}]^2 + [imf_{15}]^2 + [imf_{16}]^2\} Rg_{00}/2$$
$$[8.116]$$

Now on a cosmological scale, we can make all the *imf* factors equal. There is no reason to think the intermanifold factor will be different for the six spacetime manifolds; at the time of the Big Bang, the mass was symmetrically distributed over all six. This is only true on a cosmic scale.

We set the intermanifold factor to the average value of *imf* and remove the indices. In local clumps, these factors can vary substantially- hence the treatment of them as scalar fields as a function of spatial position. This will simplify this as follows:

$$(Rg_{00}/2)' = \{1 + 5[imf]^2\} \, Rg_{00}/2 \qquad\qquad [8.117]$$

If we set the modified constant G to :

$$G_{00}' = R_{00}' - (Rg_{00}/2)' \qquad\qquad [8.118]$$

The equation then simplifies to:

$$G_{00}' + \kappa imf \{T_{44} + T_{88} + T_{1212} + T_{1616} + T_{2020}\} = -\kappa T_{00}$$

$$[8.119]$$

Recall that the interpretation of T_{00} is the energy density. Likewise, T_{44} ,T_{88}, T_{1212} ,T_{1616} and T_{2020} are energy densities from the other manifolds. You can loosely view this as swirls and counter swirls on stress energy tensors caused by matter in different manifolds affecting adjacent common inter-manifold regions.

Viewing the stress energy tensor as the rate of flow of components of four momentum, it is as if you have eddies and currents affecting all manifolds. The movement we have identified of the quantum particles flowing through 1->2->3->4->5->6->5->4->3->2->1 is consistent with the global view of interlocked effects.

 It makes sense that if multiple spacetime manifolds are separated by intermanifold energy layers, there is a tempering effect on how much the native spacetime manifold can affect the stress energy tensor. The parallel spacetime manifolds are having an action of their own. If you think of the inter-manifold region as behaving similarly to a hydrodynamic fluid or a plasma, then the pressure exerted by the shearing effects of the other manifolds tempers the effect of our own.

You can see how equation 8.119 is remarkably close to Einstein's field equation with an additional term that looks like a cosmological constant term. Recall that Einstein had his equation with the cosmological constant term as:

$$G_{00} + \Lambda g_{00} = -\kappa T_{00} \qquad\qquad [8.120]$$

The addition of the term was ad hoc, something Einstein himself realized. But modern observations have justified adding this term in- it explains a lot about the structure of the universe. This heuristic derivation shows that, in my proposed model, the ***contribution of the stress-energy tensors from all the other spacetime manifolds gives rise to the cosmological constant term in the field equations in a very natural way.***

A more rigorous derivation would treat the "non-native" manifolds as a perturbation and the intermanifold energy layers as a repulsive force. Since the intermanifold energy layer was also the source of the vacuum energy, this provides an excellent meeting point for General Relativity and quantum field theory.

This realization made me nearly fall out of my chair- once I recovered my balance I jumped with joy. It explains a lot about why we need a cosmological constant to explain an ever-expanding universe. The equations shape up very similarly in the other spacetime manifolds:

$$G_{01}' + \kappa imf\{T_{45} + T_{89} + T_{1213} + T_{1617} + T_{2021}\} = -\kappa T_{01} \quad [8.121]$$

$$G_{02}' + \kappa imf\{T_{46} + T_{810} + T_{1214} + T_{1618} + T_{2022}\} = -\kappa T_{02} \quad [8.122]$$

$$G_{03}' + \kappa imf\{T_{47} + T_{811} + T_{1215} + T_{1619} + T_{2023}\} = -\kappa T_{03} \quad [8.123]$$

$$G_{11}' + \kappa imf\{T_{55} + T_{99} + T_{1313} + T_{1717} + T_{2121}\} = -\kappa T_{11} \quad [8.124]$$

$$G_{12}' + \kappa imf\{T_{56} + T_{910} + T_{1314} + T_{1718} + T_{2122}\} = -\kappa T_{12} \quad [8.125]$$

$$G_{13}' + \kappa imf\{T_{57} + T_{911} + T_{1315} + T_{1719} + T_{2123}\} = -\kappa T_{13} \quad [8.126]$$

$$G_{22}' + \kappa imf\{T_{66} + T_{1010} + T_{1414} + T_{1818} + T_{2222}\} = -\kappa T_{22} \quad [8.127]$$

$$G_{23}' + \kappa imf\{T_{67} + T_{1011} + T_{1415} + T_{1819} + T_{2223}\} = -\kappa T_{23} \quad [8.128]$$

$$G_{33}' + \kappa imf\{T_{77} + T_{1111} + T_{1515} + T_{1919} + T_{2323}\} = -\kappa T_{33} \quad [8.129]$$

For a rough order of magnitude calculation, note that Einstein's cosmological constant is roughly 1.19×10^{-52} m^{-2}. Einstein's constant κ is $8\pi G/c^4$ and that is roughly 2.08×10^{-43} Jm^{-1}. We can roughly estimate the cosmological intermanifold average imf as 10^{-9}.

The expansion of the universe can be ascribed to the cosmological constant, which , rather than being an ad hoc constant introduced into the equations, is a natural consequence of the fact that you have 6 interconnected spacetime manifolds with pools of intermanifold energy separating them.

V. Black Holes

Black holes absorb matter and radiation and squeeze it out of our spacetime manifold. Where this matter and radiation goes is usually deemed as a philosophical question, beyond the realm of physics. In my proposed model, the extreme gravity in one spacetime manifold creates a bridge to another spacetime manifold.

The overall space is 24 dimensional, and the black hole residing in one four-dimensional spacetime manifold creates such extreme gravity that the other end of it is connected to another four-dimensional space-time manifold. This is depicted in the figure below:

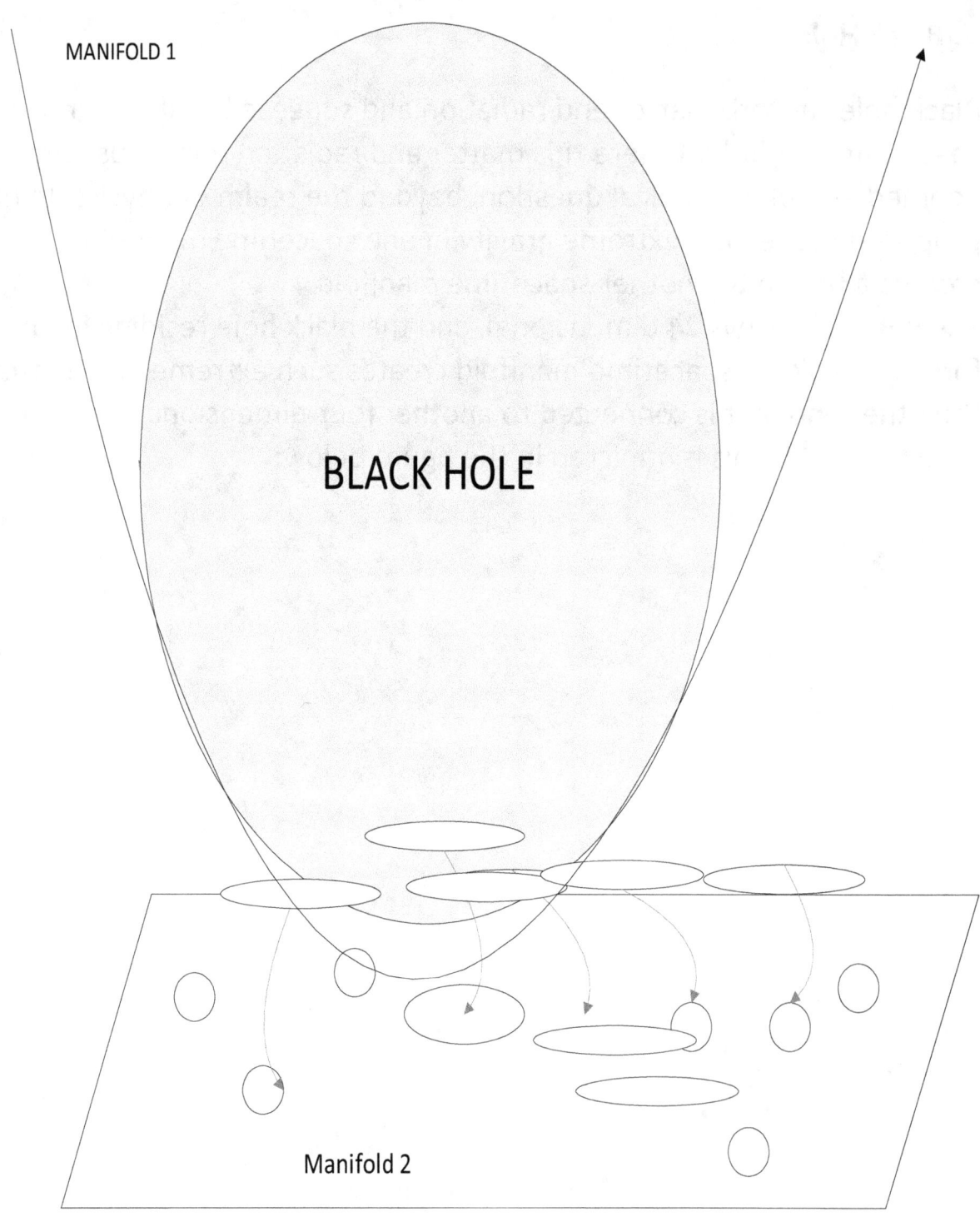

Viewed globally, the mass/energy of one spacetime manifold is being transferred to another one. The overall balance of energy and mass stays constant in the universe. That is an extremely attractive conservation law and proposition.

For example, a supermassive black hole in our galaxy (spacetime manifold 1) which is a sink for all matter and energy would become a source in spacetime manifold 2 (I chose 2 arbitrarily). Overall, the universe is balancing out these sinks and sources across all six spacetime manifolds.

The solutions and characteristics of a black hole do not change at all; it is just the interpretation of the other side being a source in an alternate spacetime manifold. This moves the output of a black hole from a metaphysical question to a concrete physical output in another spacetime manifold.

W. General Relativity vs Quantum Mechanics.

As we have discussed in prior chapters, for about a hundred years now, this has remained a source of puzzlement with various attempts to explain this fundamental difference. Both theories are experimentally accurate to an extremely high degree, so you must respect both. Yet they differ on the nature of the universe.

Our core model involves quantum folds across all the manifolds.

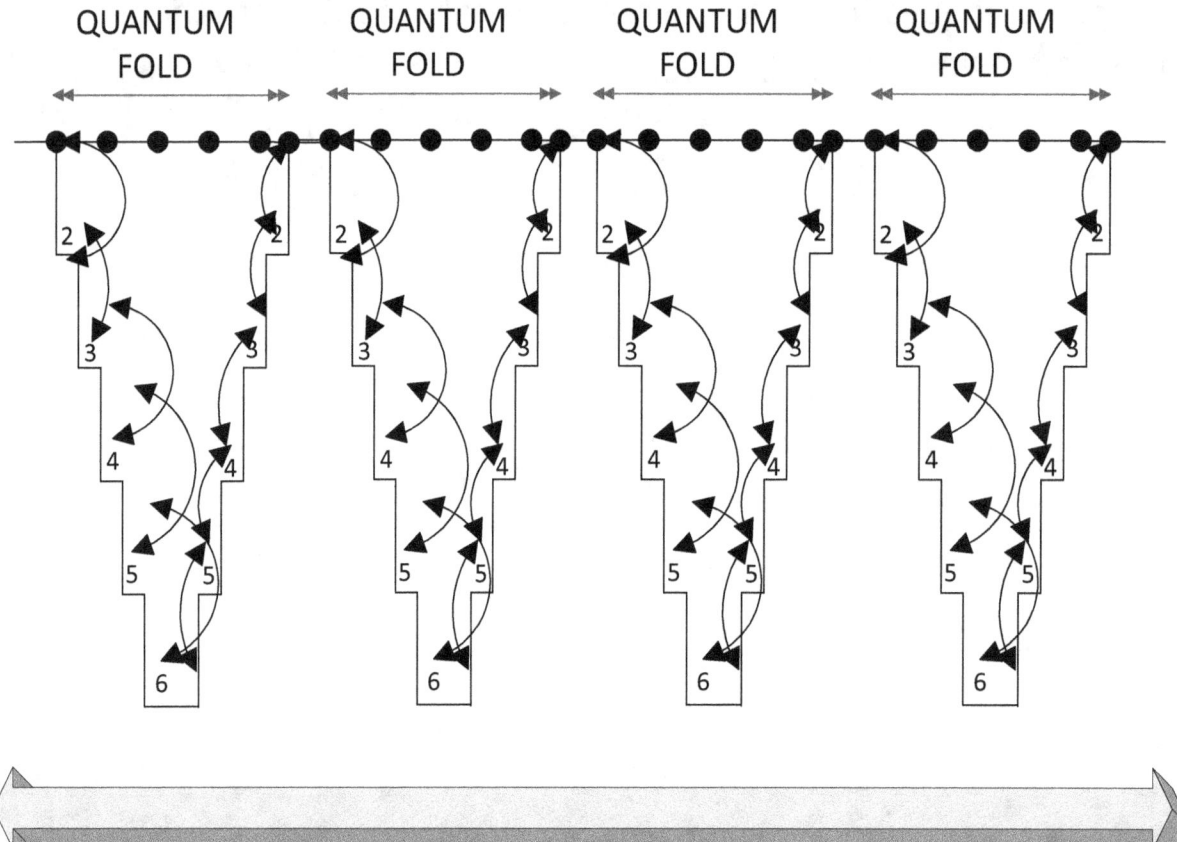

SCALE OF PARTICLE

Over the entire 24-dimensional geometry of the universe, motion is indeed continuous as shown across the quantum folds. In our specific manifold, the motion appears discontinuous.

This structure of the universe is entirely based on real manifolds with physical laws that are virtually the same in each one.

The transitions of all elementary particles in and out of the quantum folds together with their interaction with the intermanifold energy layer constitutes the vacuum energy. These particles cohere to make stable matter in each manifold. This process has a certain amount of randomness to it, and the energy left over is what quantum field theory interprets as vacuum energy.

A good analogy to get an understanding of this process is to imagine all particles undergoing transition are like individual letters in the alphabet. Sometimes the right letters combined to form smaller words, mid-sized words and large words. The words in turn form sentences, and the sentences in turn form a paragraph. The paragraphs are then put together to make a story.

The letters are akin to the elementary particles (electrons, muons, tau particles, electron neutrinos, muon neutrinos and tau neutrinos, down quarks, up quarks, strange quarks, charm quarks, bottom quarks, top quarks, photons, gluons, W-bosons, and Z-bosons) making transitions through quantum folds.

Composite particles use the elementary particles (letters) to make protons and neutrons(words). These in turn combine to make atoms(sentences). Atoms come together to make molecules (paragraphs). Finally, macroscopic objects like solids, liquids and gasses (stories) are made from the molecules (paragraphs).

As the scale of the coherent matter becomes larger and larger the quantum fold size becomes more and more irrelevant. The basic displacement of the particle becomes far larger than the quantum fold, dimming the effects of the transitions. By the time you reach macroscopic objects, it is completely irrelevant. The page that the words are written on are like the intermanifold energy layer that the manifold is imprinted on.

This is how the quantum mechanical vision and the General Relativistic vision merge in the 24-dimensional universe. As you graduate from quantum folds up to macroscopic objects, the discontinuities introduced by the quantum folds become irrelevant in each manifold. The coherence of matter in each manifold is a consequence of a sufficient number of elementary particles binding, crossing multiple quantum folds and stabilizing in that manifold. The following diagram shows how even in a limited four-dimensional spacetime manifold, the appearance of continuity is rapidly achieved:

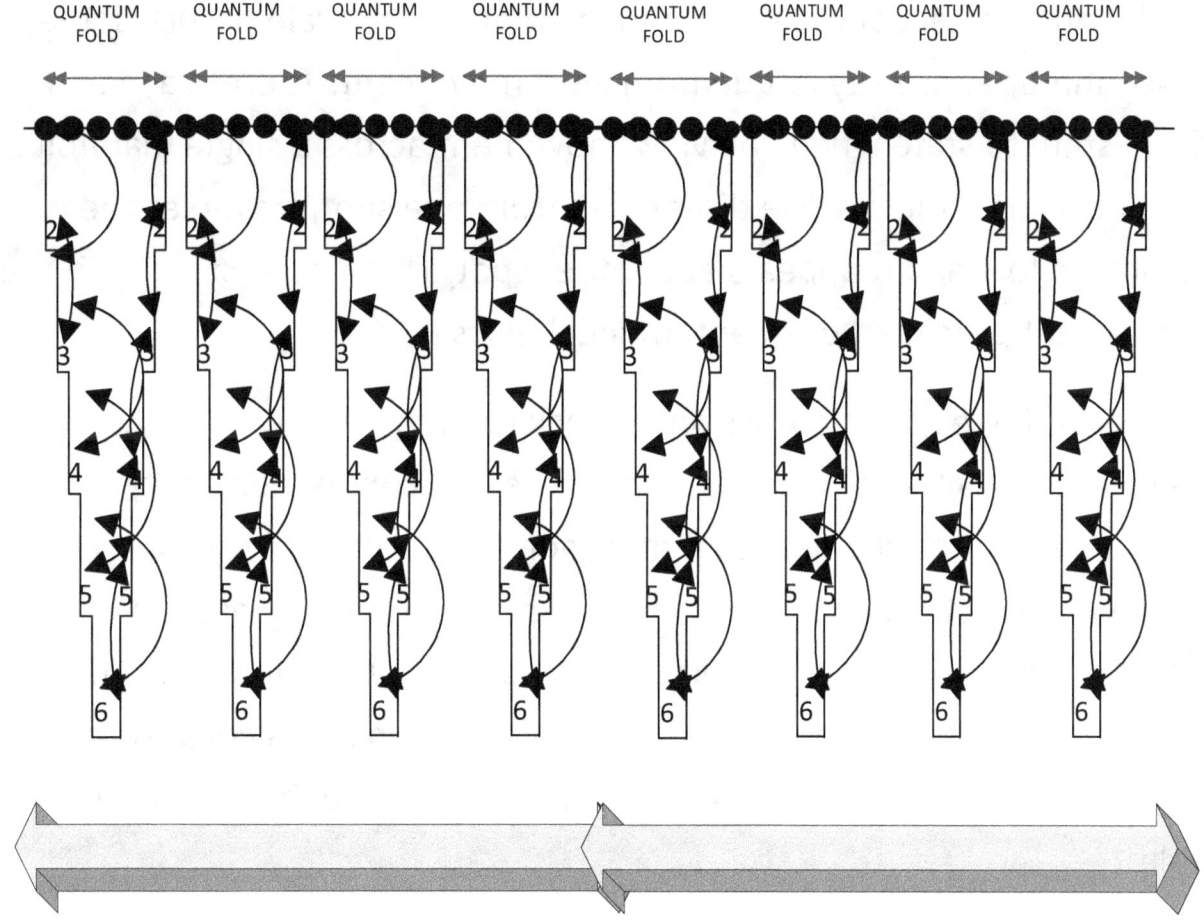

INCREASING SPACETIME SCALE

Recall that the scale of the fold (squared) itself is miniscule, on the order of the magnitude of $(\hbar^2/2)$ -Spacetime is completely continuous when viewed over 24 dimensions. When you view it within a manifold, minute quantum folds appear. These become academic for macroscopic purposes because of the coherence of matter across many folds. The view of General Relativity that spacetime is continuous within a single manifold is not entirely correct, but it is correct when viewed over the span of six manifolds.

In the other direction, as you zoom in on quantum scale particles, the discontinuities are key to quantum level movement. There is a leap from state to state when you view movement across a single manifold. It appears as if the particle disappears from one spot, traverses the quantum fold and reappears at another spot. This is consistent with the movement predicted by quantum mechanics.

However, if we adjust our vision to track it across all the six-spacetime manifolds, it will appear to be much more of a continuous entity. This puts a damper on the quantum mechanical view of a "fuzzy" particle with wavefunctions that extend everywhere in space. Consequently, the 24-dimensional universe is far more continuous than quantum mechanics suggests. The mathematics of wavefunctions collapsing into eigenstates is a representation of the traversal through the quantum fold.

Qualitatively, at least, these considerations should pave the way for a unified view of both theories without having to sacrifice a single equation known to be accurate in both.

The trick here is to get a 24-dimensional manifold to look like a four dimensional continuous one under certain restricted assumptions and using the fact that the other manifolds become apparent at smaller scales to fit in with the probabilistic interpretations that quantum mechanics provides.

It makes intuitive sense that if particles are constantly transitioning between six different manifolds at exceedingly small scales that you would have to describe their behavior using probabilities that you are seeing them in one.

From the large-scale view of the universe, General Relativity needs a perturbation introduced because of the other manifolds. These perturbations end up being part of the cosmological constant, as we saw in Chapter 8 Section U, through a heuristic, non-rigorous derivation.

From the small-scale view of the universe, Quantum Mechanics, which deals in wavefunctions, collapsing eigenstates due to measurements and probabilities, can be recast in terms of the displacements through quantum folds. The displacements involve interactions with the intermanifold energy layer, contributing to transition states.

X. History of the Universe

The history of the universe starts with a 24 dimensional initial singularity in my model. As it expanded, if you zoomed into the ultramicroscopic realm, you would find a fully 24 dimensional combined spacetime manifold, as outlined at the beginning of this chapter. This would be in the era of Planck time.

Suddenly, the universe inflates and cracks into six four dimensional spacetime manifolds, one of which we now inhabit. These are connected by quantum folds, as I postulated in the fundamental construction of my model.

The energy field that powers this inflation is the intermanifold energy. The energy squeezing through the quantum folds between manifolds appears as fluctuating energy in the vacuum. Thus, the vacuum fluctuations in standard cosmological evolution have a physical foundation. These are normally ascribed to the uncertainty principle , but we can now furnish a physical layer for these vacuum fluctuations that produce energy.

After this period, a quark/antiquark gluon plasma appears in the universe. At this stage of standard cosmology, the antimatter suddenly disappears with no explanation; in my model, the quark/antiquark gluon plasma gets distributed symmetrically between the six manifolds. A cascade effect results in matter dominance in three of them, while antimatter dominates in the other three. At this point, the electroweak force splits into electromagnetism and the weak force in each manifold.

In the penultimate phase, nucleosynthesis starts in all six manifolds. The ultimate transition phases are driven by the intermanifold energy and the matter in all manifolds affecting each other gravitationally; at this early time, the intermanifold distances are small and they affect each other intensely.

My model results in a very natural evolution for the universe with unidentifiable forces in standard cosmology identified as intermanifold energy. And unidentifiable dark matter is identified physically as matter in other manifolds. Quantum fluctuations that appear in the vacuum are the result of energy squeezing through quantum folds between manifolds. Note that explaining quantum fluctuations as the result of the uncertainty principle as in standard cosmological evolutionary theories is quite unsatisfactory; it just pushes the question back one level as to why the uncertainty principle exists. The explanation in my model provides a physical foundation.

Y. Newton's Law of Universal Gravitation

In a single spacetime manifold, as at the Big Bang, you have the usual product of masses with the inverse square law. The force between manifolds cannot be zero since we are postulating that gravity crosses the six space time manifolds. On the other hand, it cannot be the same as the usual force because it would be detectable by its very presence and effect of masses near it.

For a local manifold, the usual law applies.

$$F = G\frac{m_1 m_2}{r^2}$$
[8.130]

$$where\ G = 6.67 \times 10^{-11}\ \text{m}^3\text{kg}^{-1}\text{s}^{-2}$$

This situation is depicted in the diagram below:

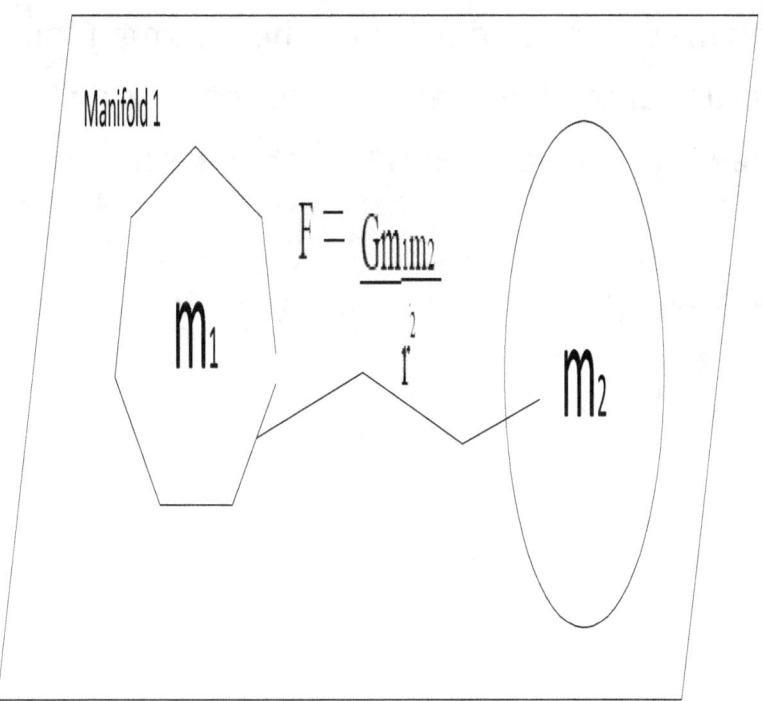

INTERMANIFOLD ENERGY

In case m_1 in manifold 1 and m_2 is in manifold 2 then the intermanifold factor comes into play. Recall that the flat space metric contains an intermanifold factor and Newton's law becomes (m_1 is always in manifold 1 in the following):

$$^2F = imf_{12} \times F \quad (m_2 \text{ is in manifold 2}) \qquad [8.131]$$

This situation is depicted in the figure below:

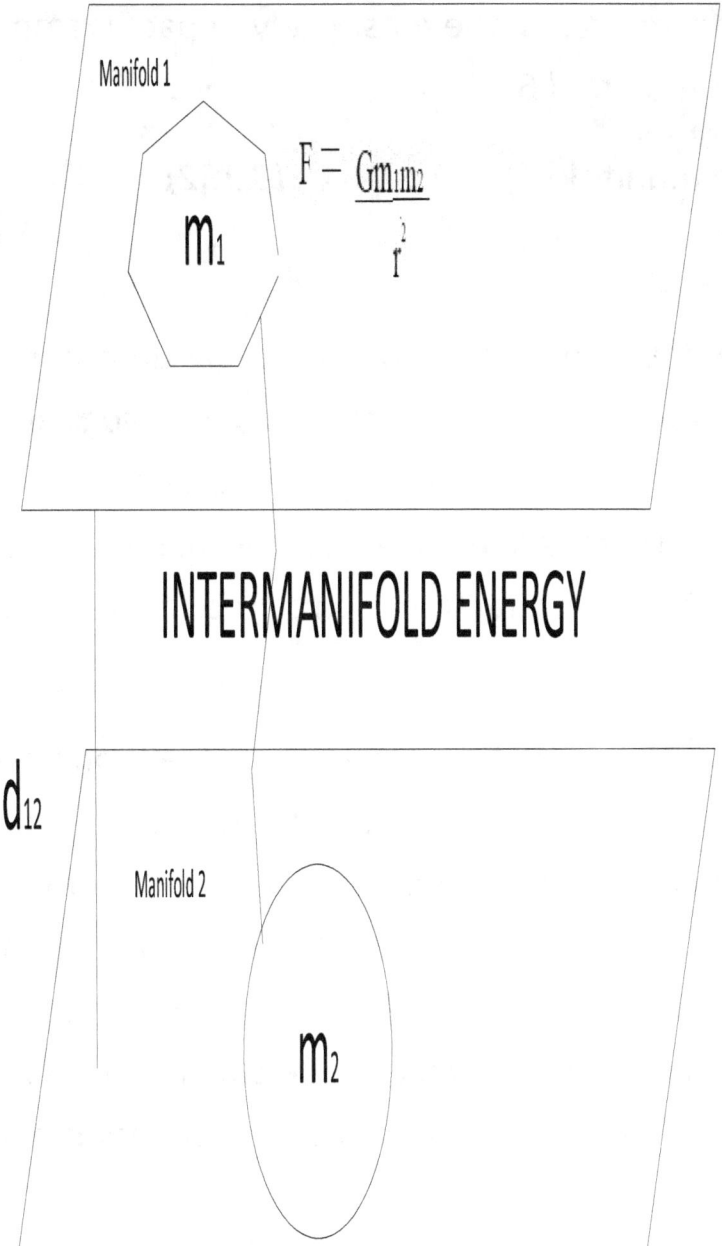

Manifold 1

$$F = \frac{Gm_1m_2}{r^2}$$

m_1

INTERMANIFOLD ENERGY

d_{12}

Manifold 2

m_2

Note that manifold 1 is adjacent to manifold 2 and manifold 6 in our model, so those will likely have the biggest impact on the first mass in manifold 1 . Although the other manifolds will theoretically impact the mass in manifold 1 in extreme cases, the most likely impact is from these two manifolds. For manifold 6,

$$^6F = imf_{16} \times F \quad (m_2 \text{ is in manifold 6}) \qquad [8.132]$$

Regardless of the source of the extra manifold mass, we determined the rough order of magnitude for imf to be 10^{-9} as a cosmological average. This would indicate that you could repeat the Cavendish experiment on Earth and if there is a mass in another manifold nearby, there could be 1 part per billion interactions compared to the norm to the mass in the apparatus.

The experiment should be set up so that essentially there is one mass only in our space time manifold, with just a stub in the second spot where the Cavendish experiment would normally have another mass. Then just assume there is another mass in another spacetime manifold "nearby" and see if the attractive force can be measured. Assume that the other mass is (effectively) orders of magnitude smaller in mass than the primary mass. This is due to the intermanifold attenuation of the gravitational force.

It would be a very delicate experiment because the experimental apparatus would be affected by an extremely small perturbation by the extramanifold mass nearby. It will be quite an experimental challenge to accurately measure this perturbation. If it works, it demonstrates in a remarkably simple way the presence of ordinary mass in another spacetime manifold. Repeating the experiment in various areas of deep space could reveal where there are concentrations of mass in an adjacent manifold, resulting in a greater skew. This would show the varying values of *imf* in my model.

Conclusions

To summarize my model neatly and bring it home, here is what the universe looks like in the vicinity of Earth:

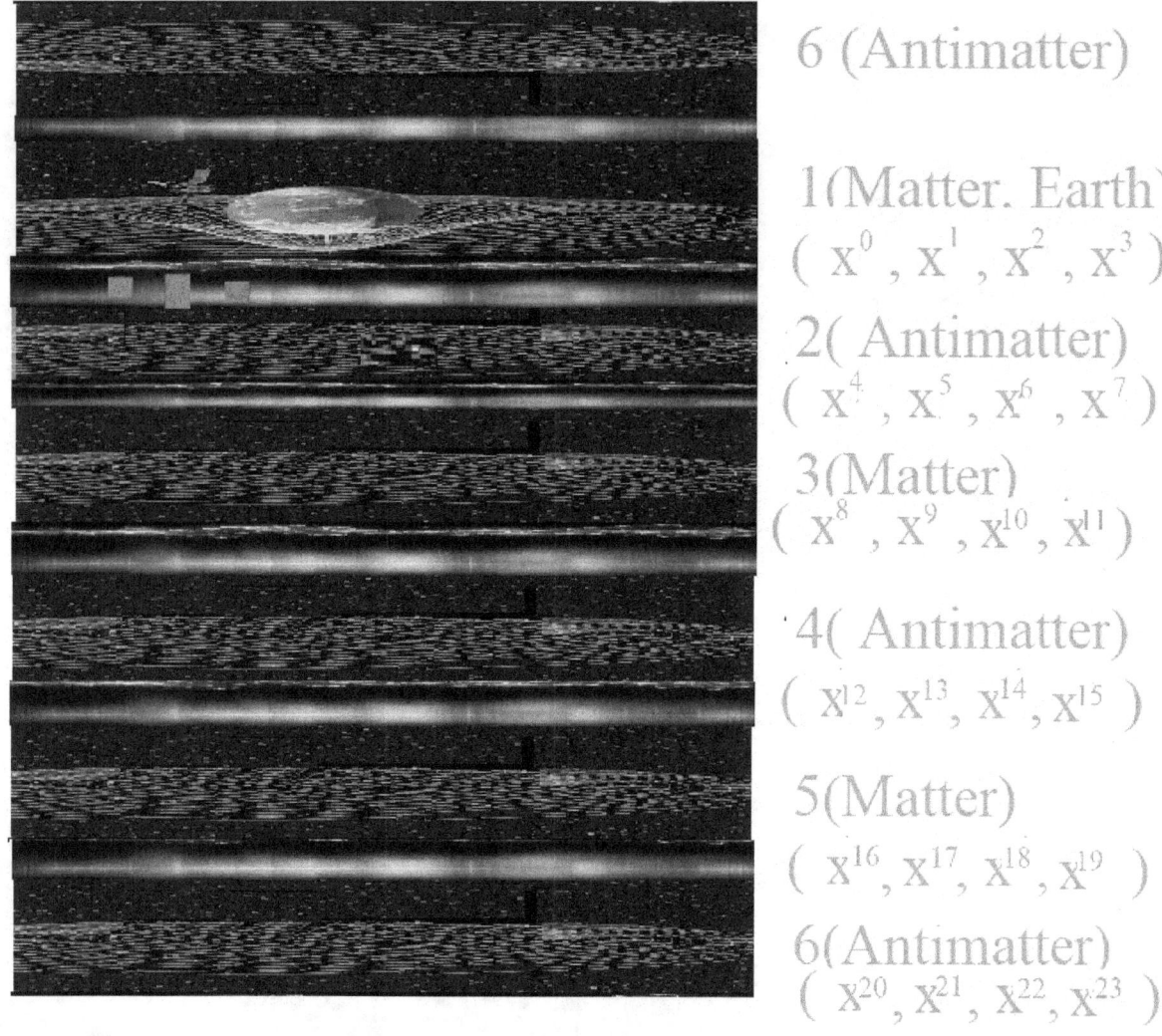

6 (Antimatter)

1(Matter. Earth)
(x^0 , x^1 , x^2 , x^3)

2(Antimatter)
(x^4 , x^5 , x^6 , x^7)

3(Matter)
$(x^8 , x^9 , x^{10} , x^{11})$

4(Antimatter)
$(x^{12} , x^{13} , x^{14} , x^{15})$

5(Matter)
$(x^{16} , x^{17} , x^{18} , x^{19})$

6(Antimatter)
$(x^{20} , x^{21} , x^{22} , x^{23})$

Our planet with a satellite is shown in manifold 1. The rest of the manifolds are shown adjacent to us. I believe a good theoretical model should provide clear, understandable answers to fundamental questions. So let us return to the basic questions that were introduced in Chapters 1-7 and see if we can provide some straightforward answers.

- ➢ What is dark energy ?
    - o The intermanifold energy
- ➢ What causes the expansion of the universe?
    - o The intermanifold energy pushing the manifolds apart since the Big Bang.
- ➢ What is the vacuum energy?
    - o The intermanifold energy, propagated through small quantum folds.
- ➢ Why is there uncertainty in quantum mechanics?
    - o Because we cannot measure physical observables in other manifolds.
- ➢ Why do we add hidden dimensions/manifolds?
    - o We do not add them, as they have always been there. They just manifest in a different form after the Big Bang.
- ➢ What causes galactic rotation anomalies?
    - o All the mass in every manifold other than our own.
- ➢ Why do we see particle and wave properties combined in nature?
    - o It is a result of being in and out of the manifold at small spacetime slices. We can only see what is in our manifold.
- ➢ How are renormalization infinities addressed?
    - o All elementary particles can be represented as waves, eliminating point particles and self-energy infinities.
- ➢ Why is the vacuum energy calculation currently so high ?
    - o You need to adjust for energy trapped in intermanifold layers. Only tiny portions of that energy squeeze through quantum folds.
- ➢ What happens to all our current laws of physics?
    - o They stay intact, in the limit of neglecting effects from other manifolds, which is true for a diverse number of situations.
- ➢ Why is there an imbalance between matter and antimatter?

- There is not. It balances out over the six manifolds.
- Why is dark matter an asymmetric quantity in the universe?
  - It is not. It is evenly distributed over all manifolds other than our own.
- Why is dark matter dark ?
  - The intermanifold energy layer inhibits photons from traversing between manifolds.
- Can you explain the proportions of dark matter and dark energy ?
  - 5% local mass *5= 25% dark matter in 5 manifolds , plus 11.67%*6= 70% intermanifold energy=dark energy.
- Why does gravity cross manifolds?
  - The repulsive intermanifold energy layer force dulls the effect of gravity across manifolds but does not eliminate it.
- Why do the other forces stop between manifolds?
  - The elementary particles of the other forces are short range and need to traverse through the folds and across manifolds for persistent activity. They are native to their manifold.
- Can you explain the cosmological constant?
  - It arises due to the perturbation created by other manifolds on our own and the energy layer between.
- How does this help with a unified field theory?
  - It combines the notion of a granular distance for spacetime (as in loop quantum gravity) with the notion of all particles having the structure of waves (similar to strings in string theory). This paves the way to borrow the best concepts from these promising theories.
- Can you calculate anything useful with this theory?
  - You can calculate the period of the outer arm of the galaxy, which matches observations.
- Can you do any experiment to support the theory?
  - The theory predicts an intermanifold factor of 1 part per billion cosmologically. Repeat the Cavendish experiment in

outer space using a stub for one end of the apparatus and look for extra-manifold mass effects on the apparatus.

➢ Can you explain the history of the universe in the framework of this model?
   o The Big Bang consists of compactified energy in 24 dimensions. The explosion cracked and split the dimensions into six 4 dimensional manifolds connected by ultramicroscopic quantum folds. The intermanifold energy drives the expansion of the universe and provides vacuum energy. Matter and antimatter are evenly distributed over the six manifolds.

➢ How does the Standard Model fit into this theory?
   o The Standard Model treats elementary particles as point particles. This theory suggests they are all wave-like. Due to the ultramicroscopic size of the quantum folds, it is a good approximation to treat elementary particles as point particles.

➢ How does this model fit in with quantum mechanics?
   o Since all elementary particles are wavelike across six manifolds, it fits in quite well with quantum mechanics. Some reformulation and reinterpretation of concepts are necessary.

➢ How does the uncertainty principle fit into this theory?
   o The granular size of the quantum fold is related to the uncertainty principle.

➢ What gives rise to mass and macroscopic objects in this model?
   o Coherence of elementary particle/waves across multiple quantum folds. Coherence occurs in each manifold giving rise to the distribution of mass in the universe. Macroscopic objects are those that cohered their many constituent wavefunctions that cross six manifolds into one manifold through stochastic processes for a huge number of waves.

- ➤ How can you explain virtual particles?
  - o Virtual particles are real particles. The intermanifold energy is the source of vacuum energy bubbling through the quantum folds. Energy "borrowed" from the vacuum is real energy.

There will be objections that the theories and speculation presented in this book lack mathematical rigor- that is no doubt true. That objection reminds me of the John Von Neumann quote: "There is no point in being precise when you don't even know what you're talking about." We need to be humble and realize we really do not even know what we are talking about when it comes to understanding the fundamental paradoxes that arise when following the basic theories of physics to their logical conclusions.

The focus for unified field theories should be on creating the conceptual framework that could at least, in principle, explain some of the deepest mysteries of physics. Theories begin with a fundamental idea- and I think the foundational ideas that are presented in this book will end up explaining many mysteries once the theory is fully developed.

There is an old directive for doctors that states: "First do no harm." It is widely assumed this is part of the Hippocratic oath. This actual phrase does not appear in the Hippocratic oath- although it is easy to argue the spirit of it is there.

Therefore, this saying serves to illustrate two of my core points. First, widely held assumptions, like those laying the foundations of theoretical physics, should be questioned ferociously. I think my model does that.

Secondly, it is important to first do no harm when you are proposing novel theories of physics. New models, conjectures and theories cannot harm the existing theories that by now explain so much of the universe brilliantly. That is why I constructed a model that respects existing laws of physics scrupulously and in the limit where you are local to a spacetime manifold becomes equivalent to all known theoretical physics.

The universe is a superposition of six distinct spacetime manifolds with quantum fractures between them. This would naturally account for missing mass without any need of exotic new theories involving parallel universes, multiple new spatial dimensions (with no accompanying time dimensions), new particles to account for mass, multiverses, etc.
We live in one universe that is a superposition of six distinct spacetime manifolds with ordinary mass distributed among them, interacting in places where we do not detect mass directly.
All the mass in the remaining spacetime manifolds is just like the mass we see around us- it is just distributed across a different space time manifold. This casts dark matter in a fundamentally different light. It is just matter that happens to be hidden in a different spacetime manifold but is just ordinary matter whose gravitational effects are tempered by intermanifold factors.
A lot of the exotic theories with particles popping in and out, infinite seas of particles annihilating each other, etc. can be replaced I think with extremely limited scale interactions between the manifolds.
This model explains in a fairly simple way why quantum mechanics must deal with probabilities and wavefunctions. At exceedingly small quantum scales, particles traverse across the six manifolds in a continual dance.
When you try to describe where anything is located or its fundamental characteristics, you are forced to use probabilities since the particle can be in another spacetime manifold for short spacetime intervals.
No modification is required to the mathematical formulation of quantum mechanics. The coefficients that describe transitions from spacetime manifolds look very much like probability coefficients found

in quantum mechanics. Cosmological phenomenon that needs extra mass as an explanation are explained by the 'hidden' mass in the extra spacetime manifolds. The geometry of the universe explains the extra mass.

Traditionally, theoretical physicists have added several spatial dimensions but have been reluctant to look at adding additional time dimensions in the form of extra spacetime manifolds; this is natural since adding dimensions in time in any form goes against some considerably basic notions of time which we all share.
However, this is largely a human-centric bias; in reality, space and time are intrinsically linked. Looking at it mathematically, I do not see any big reason for a distinction of this nature, and we should be more inclined to add spacetime in dimensional blocks. If it is legitimate to hide mass in extra spatial dimensions, surely it is at least equally legitimate to hide it in extra spacetime manifolds that happen to be weaved together and connected via microscopic gaps.
In fact, it should be more legitimate to do so, since spacetime as a combined dimension is a more fundamentally correct one. The universe is a Lego set where the basic building block is a spacetime manifold.

If relativistic effects were something within our intuitive grasp, we might have been instinctively inclined to add spacetime manifolds to the equations instead of purely spatial dimensions. I think much of what seem like bizarre phenomenon in theoretical physics will be explained very naturally in a six-way spacetime manifold setting. Quantum field theories, instead of coping with diverging infinities, could potentially replace these quantities with particles that transition between spacetime manifolds.

Modern theoretical physics is like the sections of an excellent orchestra at this point. The strings section, like General Relativity, is exquisitely beautiful and simple. The woodwind section, like Quantum Mechanics, is not easily appreciated but also beautiful and moving.

The brass and percussion sections, like the Standard Model, are more difficult for the general audience to appreciate, but nonetheless forceful and play an important role. Each section when they try to play music sound excellent- when they try to play it together, it degenerates into a cacophony.

The solution is not to fire the sections of the orchestra; a strong conductor and coordinator is required. What I am proposing is that the conductor comes in the form of a six-way spacetime manifold that came about at the time of the Big Bang.

An excellent conductor enhances the combined music of the sections. This is exactly what the multiple spacetime manifold construct does. It preserves the validity of all existing equations and adds some very plausible interpretations as to why they work.

 In addition to doing all this, it gives a clear idea of what dark matter and dark energy are and their role in our universe. The cosmological constant, which has been an excellent mechanism for explaining the universe, turns out to be a term added due to presence of the extra manifolds.

There is a hidden beauty and simplicity that is revealed with this view of the universe-ultimately, I believe a revised mathematical formulation will make this plain. This view has been hitherto unnoticed because the notion of parallel spatial dimensions marching to the tune of different chronological drummers is one that goes against all our instincts.

There is undoubtedly a speculative element in this model; most theories do begin with a speculative component. Since models to date have not been able to explain some fundamental properties of the universe, this novel approach might prompt some brilliant young physicist or philosopher of science to reexamine some fundamental assumptions and come up with an elegant new theory.

I feel intuitively that there is some fundamental truth in this model that will advance theoretical physics. As an incentive for adventurous theoretical physicists or philosophers of science out there, I am offering a $30,000 prize to any of them that wants to have fun and explore this embryonic theory. The author can be contacted at pardu71@gmail.com if you have interest in pursuing this prize.

There are only a few conditions:
- ➢ The prospective physicist must have a doctorate in physics from an accredited university. Alternatively, the prospective philosopher of science must have a doctorate in philosophy from an accredited university.
- ➢ I should be able to interview the physicist or the philosopher at the university where he/she works.
- ➢ The work should be developed so that it can be submitted to a peer reviewed journal, with my name as a coauthor. Rejections are expected, there is no point in fearing failure when you are proposing ideas that are unconventional. Persistence is required.
- ➢ I reserve the right to edit the submitted paper.
- ➢ Note that whoever works on this gets to keep the Nobel Prize in Physics they will undoubtedly win for being the person who finally produces a theory of everything that fits in with relativity, quantum mechanics and gravity!

I can provide the broad sketch of what is required:
- ➢ A 24-dimensional universe to start with, including six distinct four-dimensional spacetime manifolds.
- ➢ An explosion (the Big Bang) that results in the shearing of these manifolds, connected by quantum folds.

abandoned in its current form and replaced with real values that make sense.

➢ Whether you approach it from the relativistic viewpoint, or the quantum field theory viewpoint, the math will meet in the middle. In the former a direct sum of six distinct four-dimensional manifolds overlaps with displacement coefficients that cause a probabilistic interpretation at small scales. In the latter approach, the vacuum energy and renormalization will fix themselves at the outset due to replacement of virtual particles with real ones from other manifolds and a careful weighing of real canceling energies.

- The overall manifold is a direct sum of six individual manifolds. Each submanifold is nearly continuous, with quantum folds disrupting the continuity on microscopic scales.
- Develop a model of the quantum fold displacement for smaller particles that will explain the probabilistic view offered by quantum mechanics.
- For larger particles, the minute quantum folds becoming negligent, thus resulting in the standard smooth space time manifold view of relativity.
- Dark matter is just the matter in all other manifolds than your own. Dark energy is what is between all the 6 manifolds, which I call intermanifold energy.
- Explanations for renormalization , pinning down all the virtual particles in the explanations as real particles in manifolds other than your own.
- Natural adjustments for the artificially high quantum vacuum energy, using the concept of pooled intermanifold energy.
- Remarkably simple calculations to explain galactic orbital anomalies using extra manifold mass that influences our gravity, attenuated by an intermanifold factor.
- In terms of the math, if you want to start from the relativistic viewpoint, view our spacetime manifold as one of six and the overall universe as a direct sum of distinct submanifolds and follow where the equations take you. I feel that you will be led directly to an explanation of the cosmological constant, black holes, dark matter, and other phenomena.
- If you want to start with the quantum field theory approach, treat all hidden dimensions as extra space time dimensions (instead of extra spatial dimensions). Abandon the notion of virtual particles and treat them as real particles coming from a pool of interdimensional energy and follow where the equations take you. I feel you will get to a point where renormalization can be